THE
WEST RIDING
OF
YORKSHIRE

THE

WEST RIDING

OF

YORKSHIRE

by

BERNARD HOBSON, M.Sc., F.G.S.

Late Lecturer in Geology in the Victoria University of Manchester

With Maps, Diagrams, and Illustrations

Cambridge :

At the University Press

1921

CAMBRIDGE UNIVERSITY PRESS
Cambridge, New York, Melbourne, Madrid, Cape Town,
Singapore, São Paulo, Delhi, Mexico City

Cambridge University Press
The Edinburgh Building, Cambridge CB2 8RU, UK

Published in the United States of America by Cambridge University Press, New York

www.cambridge.org
Information on this title: www.cambridge.org/9781107657571

First published 1921
First paperback edition 2013

A catalogue record for this publication is available from the British Library

ISBN 978-1-107-65757-1 Paperback

PREFACE

THE writer is indebted to Dr H. R. Mill for the Rainfall Map and meteorological information. Mr H. Crowther, Mr J. Firth, Mr P. Haswell, Mr E. Howarth, and Mr A. Russell have furnished meteorological data. Mr C. M. Rankin and Mr F. Arnold Lees have given botanical help. Mr Harold Brodrick has revised the account of caverns and underground drainage. Mr T. H. Mottram, Dr A. Strahan, and Prof. Fearnsides have supplied some of the particulars of mineral output. Mr F. A. Darwin, Clerk to the County Council, has given details of existing administration. Mr Godfrey Bingley has given the use of his photographic negatives. The publishers of the *Dictionary of National Biography* have given permission to quote from that work the details (with a few exceptions) given in the Roll of Honour. The General Editor of the Series, Dr Guillemard, has been particularly helpful. To all of these and to the authors of the numerous books and original papers consulted, but especially to the contributors to the Victoria County History and Sir W. Boyd Dawkins, the writer expresses his thanks. He will be obliged by information of any errors.   B. H.

20 HALLAMGATE ROAD
    SHEFFIELD
    December 1920

CONTENTS

ILLUSTRATIONS

ILLUSTRATIONS

The illustrations on pp. 9, 14, 17, 19, 27, 31, 34, 35, 40, 42, 44, 102, 104, 108, 117, 119, 133, 175 are from photographs kindly supplied by Mr Godfrey Bingley; that on p. 64 from a photograph supplied by Hillabys, Ltd., Pontefract; that on p. 70 from a photograph supplied by John Crossley & Sons, Ltd. ; that on p. 74 was supplied by T. Turner & Co., Ltd. (Sheffield); that on p. 76 was supplied by Davy Bros., Ltd. ; that on p. 86 is from a photograph supplied by The Goole Times Co., Ltd. ; those on pp. 69, 97, 110, 114, 123, 140, 165, 183 are from photographs supplied by F. Frith & Co., Ltd. ; that on p. 115 from a photograph by Mr A. Hutchinson; those on pp. 118, 125, 134 from photographs by Valentine & Sons, Ltd. ; that on p. 129 from a photograph by The Photochrom Co., Ltd. ; those on pp. 21, 29 from photographs by Mr B. Hobson; those on pp. 153, 155, 160 from photographs by Emery Walker, Ltd. ; that on p. 157 is from a portrait in the Sedgwick Museum, Cambridge; that on p. 174 from a photograph by the Rev. W. Denison. The Ingleborough Section, p. 32, is reproduced by permission of Mrs Barnwell, the Folds of Craven Section, p. 32, by permission of the Geological Society of London. The Map of the Canal System, p. 143, is based on one supplied by The Aire & Calder Navigation.

ERRATA

In the Geological Map at the end of the volume, in the index of colours, d^3 represents the Carboniferous Limestone Series, including not only the Carboniferous Limestone proper but the overlying Yoredale (and Pendleside) Beds. The Bunter f^2 underlies the Keuper f. The triangular area at the western foot of Ingleborough coloured as Millstone Grit is really Ingletonian and Ordovician. The Permian Red Marls and Sandstone (e) west of Ingleton are surrounded by a small area of Coal Measures. The area near Sedbergh, marked B, is diabase, not granite.

1. County and Shire

THE counties of England are neither all alike in their origin nor in the date at which they came into existence. Some, for example Norfolk and Suffolk, are divisions of ancient kingdoms, in their case of East Anglia ; others, such as Kent, Sussex, Essex, Middlesex, and Surrey, are ancient kingdoms. Yorkshire represents the ancient Anglian kingdom of Deira. It was recognised as a shire before the Norman Conquest, and appears in Domesday Book (1086) as Eurvicescire.

The word shire is of Anglo-Saxon origin, and is usually stated to be derived from the verb *sciran* to shear, indicating a part shorn off from a larger division, but other authorities derive the word from *scír*, meaning office, charge, or administration ; hence a district governed by one or more officials. These officials, before the Norman conquest, were the sheriff (shire-reeve, from Anglo-Saxon *scir-gerefa*), and an ealdorman or earl. The word county was introduced into England by the Normans, and is derived from the Anglo-French *conté*, from Old French *conte*=a count. A county is the district originally governed by an earl or count, and the word is now used as an equivalent to shire. The origin of the name Yorkshire is almost self-evident. It is the shire of which York is the capital. The name York is derived from

the Roman name Eboracum, which became Eoforwic
of the Anglian conquerors, Ioforvik, Iorvik (pronounced
Yorvic) of the Danes, whence we get Yorick and York.

Even before the Norman conquest Yorkshire was
divided into three parts, which we now call ridings.
The word riding is derived from an Anglo-Saxon adapta-
tion of Old Norse, meaning " a third part," whence the
Early Middle English *trithing* or *triding*.   The initial *t*
afterwards became absorbed in the final *th* or *t* of the
words, North, East, and West.

The changes through which the kingdom of Deira
passed until it became Yorkshire will be alluded to
under the chapter on History.   In the reign of Henry II,
in the year 1177, when the whole kingdom was organised
into circuits of visiting judges, Yorkshire received the
territorial limits which it has retained, with slight
variation, to the present day.

2. General Characteristics.   Position and Natural Conditions

The West Riding of Yorkshire is practically an inland
region, though in Goole (on the Ouse) it possesses a seaport.   The riding is chiefly industrial and agricultural.
It is a region of great contrasts, for while in the Vale of
York it shows wide stretches of rich and level agricultural land, with a population of only moderate density,
it has one of the most densely populated industrial
regions of England in the area of the Yorkshire coalfield
to the.south-west.   In the north-west and to the west

of the coalfield lie the sheep pastures of the Pennine Chain, a district which is in great part uninhabited, except in the river valleys. These contrasts, it should be noted, are almost entirely due to the geology and physical geography of the riding.. The soil in the Vale of York consists chiefly of unconsolidated superficial deposits, such as alluvium, glacial sand, gravel, and boulder clay, and as the rainfall is moderate, the elevation below 100 feet, and the surface practically level, the conditions are favourable to agriculture.

The presence of numerous valuable coal seams is in itself sufficient to account for the large industrial population in the Yorkshire coalfield, though the water-power of the rivers has also been of value. Agriculture is carried on in this region, though it is overshadowed in importance by the manufacturing industry.

In the Pennine Chain the considerable elevation (1000 to 2400 ft.) resulting in low temperatures and a heavy rainfall, the often thin soil, and in places bare rock, are unfavourable to agriculture, but permit the growth of grass, which serves to support cattle and sheep, though large areas are covered by heather and cotton-grass moors.

3. Size. Shape. Boundaries

Before describing the West Riding, a few words may be devoted to Yorkshire as a whole. It is the largest county of England. The area of the ancient county is 3,882,328 acres, or 6066 square miles. If we could

make the county square, each side would measure
roughly 78 miles. It is thus more than four-fifths of
the size of Wales, more than twice the size of Lincolnshire, the next largest county, and nearly forty times
the size of Rutland. The greatest length of Yorkshire
is about 120 miles, whether measured from Spurn
Head to the point where Yorkshire, Westmorland, and
Durham meet, or to Carling Gill, in the Howgill Fells,
south of Tebay. The breadth from Lidgate near Oldham to Staithes near Whitby is 85 miles. Yorkshire is
bounded by Durham on the north, Westmorland on the
north-west, Lancashire on the west, Cheshire and Derbyshire on the south-west and south, Nottinghamshire
and Lincolnshire on the south-east, and the North Sea
on the east.

The area of the West Riding, as a division of the ancient
county, is 1,766,664 acres (exclusive of the city of York,
3591 acres), but the modern Administrative Riding, including the county boroughs (except York), contains
1,773,529 acres ; with York, 3730 acres, total 1,777,259
acres. This is equivalent to 2777 square miles, and if
we represent the area of the riding by a square, each
side would measure rather over 52 miles. The West
Riding alone contains 76,415 acres more than Lincolnshire, and it is more than twice as large as the average
English county. The greatest length of the riding,
from Carling Gill to Whitwell Wood, near Shireoaks, is
92 miles ; the greatest breadth, from Lidgate near
Oldham to York, is nearly 50 miles. Its shape is so
irregular as to be difficult to describe. Roughly, it may

be said to resemble a wedge lying N.W. and S.E. (the latter forming the base), and with irregular projections in the Ripon and Bowland Forest districts. The riding is bounded on the N. and N.E. by the North Riding, as far as York ; on the E. by the East Riding, from York to the junction of the Ouse and Trent ; on the S.E. by Lincolnshire, as far as Wroot, and by Nottinghamshire, as far as Shireoaks ; on the S. and S.W. by Derbyshire to Woodhead, and by Cheshire to Ogden Brook, N. of Glossop ; on the west by Lancashire, as far as Ease Gill near Kirkby Lonsdale and by Westmorland as far as Swarth Fell.

The boundaries of the West Riding are for the most part natural ones, usually following the course of a stream or river, or the line of a water-parting. Beginning in the extreme N.W. at Carling Gill, the River Lune is followed to near Sedbergh, then approximately the water-parting between Lunedale and Dentdale to Ease Gill, and that between Ease Gill and Thornton Beck to the Greeta, then the water-parting between the tributaries of the Lune, Wyre, and Loud, and those of the Hodder to near Whitewell. The River Hodder is then followed to its confluence with the Ribble, which is traced up stream to near Sawley, and thence along the Ings Beck to the water-parting between the tributaries of the Aire and Yorkshire Calder, and those of the Lancashire Calder and Roch ; then, rather unnaturally, a portion of the headwaters of the Tame (a tributary of the Mersey), is included, and the boundary then follows the streams of Head Clough and Far Small Clough, tribu-

taries of the Etherow.   It coincides with the Derwent
to Abbey Brook, where it leaves it for the water-parting
between the Derwent and the Don, then runs down the
Limb Brook and the Sheaf, up (Beauchief) Abbey Brook,
to and up the Meersbrook, down the Shire Brook, to and
up the Rother, to and up the County Dike and thence,
unnaturally including tributaries of the Trent, to Bawtry
and Epworth, and along the ancient river channel of Old
Don River to the Trent, just above its junction with the
Ouse.   The Ouse is the boundary as far as the confluence
of the Swale and Ure, and then the latter is followed,
with several remarkable deviations, due to the fact that
the wapentake of Hallikeld, between Swale and Ure,
now in the North Riding, was at the Domesday survey
(1086) in the West Riding, in which these areas have
retained their place.   From Hack Fall the boundary
follows the water-parting between the Ure on the one
hand and the Nidd, Wharfe, Ribble, and Lune tributaries on the other to the Rawthey, leaving which it
crosses Howgill Fells, coinciding with the water-parting
between the streams flowing northward to the Lune
and those running southward to the Rawthey.   Finally,
it descends Carling Gill, which was our starting-point.

4. Surface and General Features

The physical geography of the West Riding is
intimately related to its geological structure.   As
already stated, we may distinguish three main regions
—(i) that of the Pennine Chain on the west, (ii) that

of the Ouse Valley or Vale of York on the east, and (iii) that of the Coalfield Lowlands between them.

(i) The Pennine Chain, often termed the backbone of England, extends from the Tyne to near Ashbourne in Derbyshire, so that only a portion of its length is in the West Riding. Nor is the whole breadth within its borders. In the northern part of the riding, both the eastern and western slopes of the chain are included, but between Thornton and Sheffield only the eastern slope is in the riding, except in the neighbourhood of Saddleworth. The water-parting being nearer the western margin of the range, the long valleys or dales to the east have a gentle slope, while those descending into Lancashire on the west are shorter and steeper.

The Pennine Chain is far from being a simple or a continuous ridge, nor does it everywhere run in the same direction. A main ridge forms the divide between waters which reach the North Sea on the one hand, and the Irish Sea on the other. This is pierced by the Woodhead tunnel, 3 miles long, between Dunford Bridge and Woodhead, and by the Standedge tunnel, between Marsden and Saddleworth. From the main plateau-like ridge many lateral spurs run eastwards or south-eastwards for miles, often as continuous ridges, diminishing gradually in height. These are dissected by small streams, so that we may compare the chain in plan to a bracken frond. North of the latitude of Skipton there are two main ridges, which meet near Dodd Fell, firstly that forming the water-parting between the Ribble and Wharfe, secondly that separating the Wharfe and Nidd.

There is also a subordinate ridge in Dallowgill Moor, east of the Nidd.

The chief gap in the chain within the riding is the Aire Gap, the greatest height of which is about 500 ft. The Calder Gap, once the overflow channel of great glacial lakes, is just under 775 ft. The gorge is very narrow, and is 400 ft. deep on the watershed. Todmorden lies in it. The two gaps just mentioned afford facilities to canals and railways crossing the chain. The Wharfe Gap is a continuous valley at about 1250 ft. at the head of the Wharfe and Cam Beck, thus connecting Ribblesdale and Wharfedale. These narrow gaps do not however, destroy the essential continuity of the range. Between Woodhead and Todmorden the direction of the line of water-parting and main ridge is almost S.E. to N.W., then it runs N. and N.E. towards Keighley, N.W. towards Thornton, and then almost due N. to Penyghent and Dodd Fell. The water-parting is thus many miles farther west in the north of the riding than in the south.

The general height of the chain in that part of the riding to the south of Settle and Grassington is from 1000 to 1500 ft., but east of the Derwent, Margery Hill reaches 1793 ft., and north of Woodhead, Black Hill attains 1909 ft. Between Clowes Moss, near Delph, and Skipton, no height is met with exceeding 1573 ft. on the slopes of Boulsworth Hill. North of Settle and Grassington, however, large areas reach to between 1500 and 2000 ft., and fourteen hills are of still greater height. Thirteen of these are alike in geological structure, and exceed 2000 ft. in height. They consist of almost horizontal beds of Great Scar Limestone at the base, surmounted by Yoredale shales, grits, and limestones, and capped by Millstone Grit. They are Swarth Fell (2235 ft.), on the boundary of the North Riding ; Baugh (Bow) Fell (2216 ft.), north of

Ingleborough from Ravenscar

Garsdale ; Dent Crag (2250 ft.), at the junction of Lancashire, Westmorland, and Yorkshire ; Whernside (2414 ft.), the loftiest in the riding, south of Dentdale ; Ingleborough (2373 ft.), south of Whernside ; Penyghent (2273 ft.), east of Ingleborough ; Widdale Fell (2203 ft.), south-east of Baugh Fell ; Fountains Fell (2191 ft.), and Darnbrook Fell (2048 ft.), south-east of Penyghent ;

Birks Fell (2001 ft.), between Littondale and Wharfedale ; Middle Tongue (2109 ft.), north of Birks Fell ;
Buckden Pike (2302 ft.), north of Kettlewell ; and
Great Whernside (2310 ft.), east north-east of Kettlewell. Lastly we have the Howgill Fells, consisting of
very ancient (Ordovician and Silurian) rocks. They
occupy a triangular area in the extreme north-west,
between the Lune and the Rawthey, and rise to 2220
ft. in the Calf. The hills of Bowland Forest, south-west
of Settle, reach heights of 1300 to 1784 ft.

There is hardly a hill in the riding that can properly
be described as a peak. As a rule the summits are
rounded and often form long, slightly undulating, or
flat-topped, broad ridges, not sufficiently detached from
their neighbours to show much individuality. The
summits are generally covered with grass, cotton-grass
(*Eriophorum*), or ling.

(ii) Our second region is the Vale of York. In the
strictest sense its boundaries within the riding are
formed by the Ouse on the east and the eastern margin
of the Magnesian Limestone on the west. It is therefore
the area coloured as Alluvium on the geological map.
This area, as previously mentioned, is practically level,
and less than 100 ft. above the sea. The Ure and
Ouse gradually leave the western margin of the alluvial
plain, and flow south-eastward, so as to approach more
nearly to its middle. The result of this is that the width
of the alluvial flat within the riding, while only a few
hundred yards at Ripon, attains 15 miles at Goole,
and 20 at Adlingfleet.

It is more convenient, however, to include in the Vale of York the Triassic Bunter and Keuper beds, which are mostly covered by superficial glacial deposits, and attain a height of 223 ft. at Allerton Park, south of Boroughbridge. We may also include the long narrow strip of Magnesian Limestone, which, partly covered with boulder clay, sand, and gravel, dips gently eastward towards the alluvium, while its western margin usually forms a steep escarpment. The Vale of York may also be held to embrace the lower part of the basin of the Don below Conisbrough, and the small area draining towards the Trent. Hatfield Chase, Thorne Waste, and Goole Moor, south-east of the Don, are extensive peat-bogs, but have been artificially drained, the first by the Dutchman, Cornelius Vermuyden, in the reign of Charles I. As previously mentioned, the Vale of York is given up to agriculture, and thus contrasts strongly with the wild and almost uninhabited moorland of the Pennine Chain.

(iii) We may now pass to our third main region—the Coalfield Lowlands, which lie between the Vale of York on the east and the Pennine Chain on the west. Their eastern boundary is easily seen on the geological map ; it is the western margin of the Magnesian Limestone (coloured purple on the map), but their western margin is less easily defined because the Pennine Chain merges gradually into the lower ground, and any delimitation must be more or less arbitrary. We may, however, take the 400 ft. contour line, which is shown on the physical map, as a western boundary. This country forms neither a level plain, like the alluvium of the Vale

of York, nor a mountainous district, like the crests of
the Pennines.   It is an undulating region with hills 200
to 300 ft. in height, enough to relieve the landscape from
monotony without constituting formidable barriers to
communication or agriculture.   The main rivers, the
Ure, the Nidd, the Wharfe, the Aire, and the Don, flow
eastwards to the Ouse, through narrow and deep gorges
in the Magnesian Limestone.   In this region the greatest
cities and towns of the riding are situated, chiefly on
account of the coal which underlies the area.   Sheffield
lies close to the Pennine Chain ; Rotherham, Barnsley,
Wakefield, Dewsbury, and Leeds are somewhat similarly
placed.   Huddersfield, Halifax, and Bradford lie in the
valleys between the spurs of the Pennines, at or slightly
above the 400 ft. contour line.   In the Coalfield Lowlands
coal-mining, manufacturing industry, and agriculture
afford abundant occupation to the inhabitants.

5. Watersheds.   Rivers.   Lakes

The great main water-parting of England either runs
through, or forms the western boundary of, the West
Riding.   Beginning at 1877 ft. on the slopes of Dodd
Fell, it immediately sinks to little more than 1250 ft.
between the headwaters of the Wharfe and those of the
Cam Beck.   It crosses the flat, swampy ground of this
continuous valley or gap in the Pennine Chain, runs
southward to Penyghent and Hellifield (where it crosses
the Aire Gap), and passes slightly to the west of Thornton.
Thence it forms the county boundary over Boulsworth

Hill to the Calder Gap, then runs along Blackstone Edge, leaving which it cuts across a corner of the riding on Standedge, passes on to Black Hill and the ridge above Woodhead tunnel, and at Featherbed Moss finally leaves the riding.

Of the whole area of the West Riding, about 2200 square miles, or 79 per cent, drain to the Ouse and its tributaries, the remaining 577 square miles to the Lune, Ribble, Mersey, and Trent.

The Ouse, the most important river of Yorkshire, is navigable throughout its length of 63 miles. It is formed by the junction of the Ure and Swale, a little east of Boroughbridge, and flows south-east through a wide, practically level, alluvial plain. Some 9 miles from its origin is Nun Monkton and the confluence of the Nidd, and about 8 miles further the city of York is reached. Two miles and a half below York is Bishopthorpe, with the palace of the Archbishop of York, and a little further Naburn Lock, up to which the river is tidal. About a mile north of Cawood is the confluence of the Wharfe, nine miles below which Selby is passed. Between Selby and Goole the Derwent and the Aire join the Ouse, the former on the left, the Aire on the right bank, their mouths being 3 miles apart. Eight or nine miles beyond Goole the Trent joins the Ouse to form the Humber.

The Ure (Yore) rises in the North Riding, on the east flank of a continuous valley, connecting at 1194 ft. with that of the Eden. It flows down Wensley Dale (Yoredale) past Hawes, Askrigg, Aysgarth, Wensley, Middleham, Jervaulx Abbey, and Masham, and begins to form

the boundary of the West Riding at Hack Fall, a waterfall on a tributary stream. The Ure here in a deep
gorge, forms a horse-shoe loop in which smooth reaches,
overhung with splendid trees, alternate with others
where the river, foaming like a torrent, flows over moss-

The Ure at West Tanfield

grown rocks. Some two miles from Hack Fall is West
Tanfield, below which the Magnesian Limestone is
traversed in a valley bordered by river terraces ; four
miles beyond is Ripon, and some twelve miles lower the
confluence with the Swale. The total length of the Ure
is 61 miles, and the area of that portion of its basin
which lies in the West Riding is 110 square miles.

The Nidd, whose valley is called Nidderdale, rises on Great Whernside, and after a short course of 1½ miles, flows into three reservoirs of the Bradford Corporation, in all 2½ miles long. A couple of miles beyond it sinks underground, flows through a cave called Manchester Hole, and does not reappear until, after a subterranean course of 2 miles, it issues from the caverns called Nidd Heads, just beneath the high road at Lofthouse. The natural river bed below Manchester Hole is usually dry for a quarter of a mile as far as Goyden Pot, a cavern on the left bank, in which the river is swallowed if sufficiently full of water to get beyond Manchester Hole. Just below the entrance of Goyden Pot a tributary stream enters the dry river bed, and, joined by others, flows down to Lofthouse, where, below the waterfall of Lofthouse Foss, it is swallowed up in a fissure. The remainder of the bed of the Nidd is often dry until it is joined, a quarter of a mile below Lofthouse, by How Stean Beck, which forms a beautiful gorge in the limestone towards its lower end. Blayshaw Gill, which joins the Nidd at its confluence with How Stean Beck, has also much of its water swallowed. Ten miles from its source the Nidd enters Gouthwaite Reservoir, 2 miles long, and passes Pateley Bridge to reach Nidd Viaduct (N.E.R.) where a deep, narrow, and beautifully wooded gorge begins, in which Knaresborough lies. After 2 miles the gorge widens, and Cowthorpe, famous for its oak, is reached ; thence such are the windings of the stream that it takes 12 miles to reach Nun Monkton, its point of confluence with the Ouse, distant only 6 miles as the crow flies. The Nidd is not navigable. Its length is 53 miles, and the area of its basin about 254 square miles.

The Wharfe rises under the name of Oughtershaw Beck in the continuous valley of Wharfe Gap, at the foot of Dodd Fell. Some four miles from its source it is crossed by a bridge at Deepdale, just below which, in dry seasons all, and in flood time some, of its water is swallowed in the Great Scar Limestone, to reappear from a cave in the right bank, two or three hundred yards lower down. Passing Buckden, in a beautifully wooded part of the valley—which above this point is called Langstrothdale and below it Wharfedale—Kettlewell is reached, which lies among bare limestone scars. Two miles below, the river Skirfare, 12 miles long, joins on the right. In the summer of 1915 the Skirfare being low, the whole of it was swallowed in the limestone a little south of the stone bridge which crosses it near Hesleden, and its bed was dry for 2½ miles, until a tributary joined on the left. Below the confluence of the Skirfare, Kilnsey Crag, an overhanging crag of Great Scar Limestone, 165 feet high, occurs on the right. At Ghaistrills Strid the river flows through a channel only 5 feet 6 inches wide, and afterwards forms the Linton Falls. The three miles below Barden Bridge to Bolton Priory (Abbey) include perhaps the most beautiful river scenery in Yorkshire, the Strid, 1 mile below Barden Bridge, being the narrowest part of the channel. After Ilkley and Otley, Wetherby is passed, below which is a very beautifully wooded rectangular entrenched meander, a

The Wharfe, Bolton Woods

B

mile long, in the Magnesian Limestone. Boston Spa
and Tadcaster succeed and, at 74 miles from the source,
the confluence with the Ouse is reached near Cawood.
The area of the basin of the Wharfe is 408 square
miles.

The Aire has its source in a stream which, after a
short course of about a mile, flows into Malham Tarn.
This, the largest lake in the riding, lies at a height of
1246 feet, has an area of about 150 acres, and is nowhere
more than 14 feet deep. Malham Tarn Water, issuing
from the tarn, is swallowed in the limestone in less than
half a mile, and reappears at Aire Head Springs, 2 miles
lower down. A dry limestone valley extends from Malham Tarn Water Sinks to the top of the grand vertical
limestone escarpment of Malham Cove, 285 feet high.
The water issuing at the foot of the cove is derived from
the Smelt Mill stream and the area west of the dry
valley. At 4½ miles from its source the Aire is joined by
Gordale Beck, which rises north-east of Malham Tarn,
and flows down a limestone gorge to Gordale Scar, 300
feet high, perhaps the finest limestone cliffs in the West
Riding. Later, it receives the stream from Eshton
Tarn, which is now less than a mile in circumference,
though formerly larger. Passing Gargrave, Bingley,
Shipley, and Kirkstall Abbey, it reaches Leeds, after a
course of 47 miles. Near Castleford it is joined by the
Calder, and after flowing by Ferrybridge and Snaith
it joins the Ouse near Airmyn, 85 miles from its source.
Its basin, exclusive of that of the Calder, has an area of
504 square miles.

Gordale Scar

The Calder (the name of several English rivers) rises
in the deep and narrow Calder Gap in the Pennine Chain,
and by the time it reaches Todmorden, 4 miles from its
source, is much polluted. Passing Hebden Bridge, it
emerges from the higher moorlands at Sowerby Bridge,
flows through the populous regions of Brighouse, Mirfield,
Dewsbury, and Wakefield, and after a course of about 50
miles joins the Aire at Castleford. The area of the
Calder basin is 361 square miles.

The Don rises at 1500 feet on Grains Moss, close to
the Cheshire border, and flows through two reservoirs to
Dunford Bridge. After a course of 10 miles, it reaches
Penistone, and 6 miles further Wortley Station, between
which and Oughtibridge, a distance of 5 miles, the river
flows through the beautiful Wharncliffe Woods, at the
foot of the grand escarpment of Wharncliffe Crags. At
Sheffield, at the confluence of the Sheaf, the Don makes
an abrupt right-angled bend, and thenceforward flows
north-eastward. At Rotherham the Rother joins it ;
at Conisbrough the Dearne, below which the river flows
through a beautiful wooded gorge in the Magnesian
Limestone to Doncaster. Below Thorne the course of
the river has repeatedly been artificially changed. At
first it probably flowed by Crowle to Adlingfleet,and was
" diverted at various times to the Trent at Keadby, the
Aire at Snaith, and finally by the Dutch River to the
Ouse at Goole." The total length of the Don is 67
miles, and the area of that part of its basin within the
riding is 564 square miles. The Firbeck Dike and
Anston Brook join the Ryton, a tributary of the Idle,

which in its turn joins the Trent.   The Torne, rising south of Tickhill, is also a tributary of the Trent.   The area draining to the Trent, including 7 square miles, to the Derbyshire Derwent, is 150 square miles.

The Ribble rises on Wold Fell, and after a course of 9

The Ribble below Gisburn Bridge

miles, reaches Horton in Ribblesdale.   At Great Stainforth, it forms the waterfall called Stainforth Force, and directly afterwards is joined by the Cowside Beck, on which is Catterick Force, a beautiful waterfall of 60 feet. Settle is passed, and 13 miles further Gisburn Bridge, for 3 miles below which there is a wooded gorge, affording most beautiful scenery.   After a course of 40 miles the Ribble is joined at the boundary by the Hodder, and leaves the riding. The latter stream, which rises on Great Harlow, is also notable for its charming scenery. The area within the riding drained by the Ribble and the Hodder is 225 square miles.

The Lune, rising on Howgill Fells, after flowing for 13 miles, begins to form the West Riding boundary at Carling Gill, and continues to do so for 6½ miles, when it is joined by the Rawthey, which comes in from the north-east, it source being on Baugh Fell. The Greeta, on which is Ingleton, and the Wenning, on which lie High and Lower Bentham, are tributaries of the Lune, which altogether drains 178 square miles in the riding.

The only lakes in the riding besides Malham Tarn and Eshton Tarn, are a few very small tarns on Baugh Fell, Widdale Fell, Whernside, Middle Tongue, Fountains Fell, and Birks Fell. Fleet Moss Tarn and Oughtershaw Tarn are near the source of the Wharfe, and Newhouses Tarn lies 1½ miles north of Horton in Ribblesdale.

6. Geology

The existing physical features of a district are chiefly due to the nature and arrangement of the rocks, of which its mountains, valleys and plains are composed. The fossils contained in the strata not only indicate their relative age but throw light on the geography of the past. The rocks supply building materials. The useful minerals they contain help to determine the occupations of the inhabitants.

		Western Type of Permian.	Eastern Type of Permian.
QUATERNARY OR POST-TERTIARY	RECENT	Alluvium, River Terraces, Peat Depos's, and Calcareous Tufa.	
	PLEISTOCENE	Mammaliferous Cave Deposits. Glacial Deposits, Post-Glacial Deposits?	
SECONDARY OR MESOZOIC	TRIASSIC	Keuper Sandstone. Bunter Sandstone.	

		Western Type of Permian.	Eastern Type of Permian.
	PERMIAN	Upper Breccias and Sandstone. Lower Breccia.	Upper Marl. Upper Magnesian Limestone Middle Marl. Lower Magnesian Limestone. Lower Marls, Marl Slates, or Yellow Sands.
	CARBONIFEROUS	Coal Measures. Millstone Grit. Yoredale and Pendleside Series. Carboniferous Limestone. Basement Conglomerate.	
	OLD RED SANDSTONE?	Conglomerate.	

		Howgill Fells.	Ingleborough District.
PRIMARY OR PALAEOZOIC.	SILURIAN	Kirkby Moor Flags. Bannisdale Slates, Coniston Grits, { including *inter alia*, Winder Grit. Helm Knot Sandstone. Upper Coniston Flags. Middle Coniston Flags. Lower Coniston Flags. Pale Slates (of Tarannon Age), Graptolitic Mudstone (=Stockdale Shale) with Spengill Limestone Paste Rock Conglomerate. Coniston Limestone Series.	Grfts of Rough Lands. Studfold Sandstone. Horton Flags. Austwick Grit. Austwick Flags. Pale Slates (and Red Shale), Graptolitic Mudstone with Spengill Limestone. Conglomerate or Calcareous Grit. Coniston Limestone Series.
	ORDOVICIAN		
	PRE-CAMBRIAN?	Ingletonian Grits, Slates, and Conglomerate.	

Geological Formations represented in the West Riding

The accompanying table shows the geological formations represented in the West Riding. It should be mentioned that between the end of the Triassic period and the beginning of the Pleistocene, there was a vast interval of time, not represented by existing deposits within the riding.

INGLETONIAN

The oldest known rocks of the West Riding occur as small inliers (*i.e.* isolated areas of older rocks, completely surrounded by newer strata,) near Ingleton. The chief area is in the valley of the Dale Beck, and measures about 2 miles by ½ mile, but other exposures occur in the valley of the Ribble, near Horton-in-Ribblesdale, and in Crummackdale. The rocks are chiefly greenish or bluish grits, slates, and conglomerate. The grits and slates are composed of the same materials, but in the slates it is finer in texture, and they are well cleaved. The rocks are highly inclined, nearly vertical. No fossils have been found in them. By the Geological Survey they were formerly called Lower Silurian (Ordovician), from their supposed resemblance to rocks of that age in the Lake District, but by some they are held to be Pre-Cambrian.

ORDOVICIAN

Near Ingleton these rocks are represented by blue Coniston limestone and calcareous shales or mudstones (Ashgill shales), which overlie the Ingletonian in the localities already mentioned, and occur in Jenkin Beck, Clapham Beck, and Austwick Beck. On the eastern

side of the Howgill Fells two inliers of the same series
occur. Similar Ordovician rocks occur at Helm Gill
and Gawthrop near Dent. In these Ordovician rocks
numerous fossils are found, so that their geological age
is not doubtful.

SILURIAN

Silurian rocks are not exposed in the valley of the Dale
Beck, but in the inliers of Crummackdale, Ribblesdale,
and south of Malham Tarn, also in the Howgill Fells and
Dentdale. They overlie the Ordovician, and consist in
both areas of strata having a considerable thickness—
over 3400 feet in the Ingleton district, and perhaps
10,000 feet in the Howgill Fells, where their soil supports
only pasture for sheep. Their sub-divisions are given
in the table. It is evident that the Ingletonian, and
possibly the Ordovician, rocks were being eroded when
the Silurian basement conglomerate was deposited in
the Ingleton area, for it contains fragments of Ingletonian and perhaps of Ordovician rocks. At the close
of Silurian times the Ingletonian, Ordovician, and
Silurian strata, which were originally deposited as horizontal beds, were subjected to great lateral pressure,
which caused them to be thrown into folds—close folds
in the case of the Ingletonian, but more open in the
newer rocks. Not only were the rocks folded, but they
were elevated above sea-level, and subjected to erosion
by rain and rivers for a vast period of time, during
which Devonian rocks were formed elsewhere. They
then sank again beneath the sea, and as they subsided,

were worn down into a plain of marine denudation, with differences of level of several hundred feet, upon the uneven surface of which Carboniferous strata were deposited unconformably, *e.g.* at Arco Wood Quarry, Horton in Ribblesdale, and Thornton Force, near Ingleton.

CARBONIFEROUS

If we disregard superficial glacial deposits, the Carboniferous rocks occupy a much larger area than the combined area of all the rest of the rocks of the riding. There are remarkable differences between those north of the Craven faults and those south of them, in illustration of which Mr Tiddeman drew up the following table :—

Southern or Bowland Type.	Feet.		Feet.	Northern or Yoredale Type.
Millstone Grits .	3900			Millstone Grits.
Bowland Shales .	300–1000	The	400–900	
Pendleside Grits (inconstant) . .	0–250	Great		Yoredale Series.
Pendleside Limestone . . .	0–400	Craven		The Carboniferous
Shales with Limestone . . .	2500	Faults.	400–800	Limestone with Conglomerates at
Clitheroe Limestone[1]	3500 No base.			the base.

Beginning with the northern type, which is the simplest, we have at the base a conglomerate or pudding stone, which is of two types ; the first is usually dull red to chocolate in colour, and consists chiefly of pebbles of Silurian grit, Ordovician limestone, and igneous rocks. It is by some considered as of Old Red Sandstone

[1] According to Dr W. Hind, the beds between the Clitheroe Limestone and the Millstone Grit are only about 2300 feet thick,

Thornton Force, near Ingleton (*Copyright by Mr G. Bingley*)

Age. The second type is usually grey, consists chiefly of quartz pebbles, and undoubtedly belongs to the Carboniferous. These conglomerates occur for 3 miles E.N.E. of Sedbergh, also in Ingleton Dale, etc. They are the result of the erosion of the Pre-Carboniferous land, partly by rivers, partly by the sea.

The Carboniferous Limestone. This is a purely marine deposit, as is shown by its numerous fossils, and was laid down in a landlocked sea, communicating with the open ocean. It is usually a greyish limestone, often weathering almost white, and, together with the overlying Yoredale Series and Millstone Grit, lies almost horizontally. It is termed the Great Scar Limestone, from the fact that it often forms long lines of vertical "scars" or cliff-like escarpments, frequently set in a vertical series like giant steps, the lower escarpment projecting farthest into the valley, and each succeeding escarpment being recessed farther back. Examples near Ingleton are Keld Head Scar, Twistleton Scar, and on Ingleborough, Raven Scar (see view, p. 9).

Another striking feature of the Carboniferous Limestone is the frequent development of natural limestone pavements, locally called "helks," "clints," or "grykes," where its upper surface is exposed and denuded of soil. The carbonic acid in the rain has chemically corroded the rock along joints, so that ridges of limestone, from 6 inches to 5 or more feet high, stand out in relief, and are separated one from another by spaces of depth equal to the height of the ridges, and from a few inches to several feet across. In these fissures ferns and wild

flowers often flourish. An excellent example of this
occurs at the top of the Malham Cove escarpment.
Lastly, caverns, swallow-holes, or " pots," and underground watercourses—to be described in a subsequent
chapter—are abundant in the limestone. The soil of

Limestone " pavement " above Malham Cove

the Carboniferous Limestone is generally thin, but it
produces, even on the higher hills, a grass which is sweet
and nutritious.

The Yoredale Series overlies the Carboniferous Limestone, and takes its name from the valley of the Ure. It
consists of a series of shales and sandstones, interbedded
with limestones, seven beds of which have received

separate names, but they are frequently not all present. Their fossils are chiefly marine, though coal occurs.

The Millstone Grits by no means consist exclusively of grit and sandstone, but largely of shales, alternating with the grits and a few thin beds of coal. They were chiefly deposited in fresh water, but south of the Craven faults some beds with marine fossils occur. The hard grits usually form bold escarpments, for example, that forming the flat summit and fortress-like cliffs of Ingleborough. North of the Craven faults the Millstone Grit only occurs in small isolated outliers, and in these only the lower beds of the Millstone Grit have survived ; the main mass occurs at and east of Great Whernside, and in the Pennine Chain, east and south of Skipton.

In Chapter IV thirteen hills of above 2000 feet in height, and of uniform geological structure, which occur to the north of the Craven faults, were mentioned. A few words may be devoted to an explanation of their origin. The area in which they occur consists of an uplifted block of Carboniferous rocks. Over all this area east of the Dent fault, and north of the Craven faults, the Carboniferous Limestone, together with the overlying Yoredale Rocks and Millstone Grit, once formed a continuous unbroken sheet of strata. On this surface rain and rivers, aided by heat and frost (and, at times, glaciers) have ceaselessly been at work, since it was last raised above the sea, carving out channels and valleys. In this way the separate hills mentioned, and many more of less height, have been left in relief, until they now form mountains of circumdenudation, in which each bed was once continuous with a corresponding bed in an adjoining hill. (See section, p. 32.)

We may now turn to the Southern or Bowland type of Carboniferous, which occurs in the ' Craven Lowlands (not devoid of hills), south of the Craven faults. Here we find the strata much thicker ; for instance, the

Folded Carboniferous Limestone, Hambleton Quarry, near Bolton Abbey

Carboniferous Limestone, at a point south-west of Gisburn, is 3250 feet thick, without its base being visible, so that its total thickness must be even greater. No entirely satisfactory explanation of this great and rapid increase of thickness has been offered. Another remarkable difference between the Carboniferous rocks

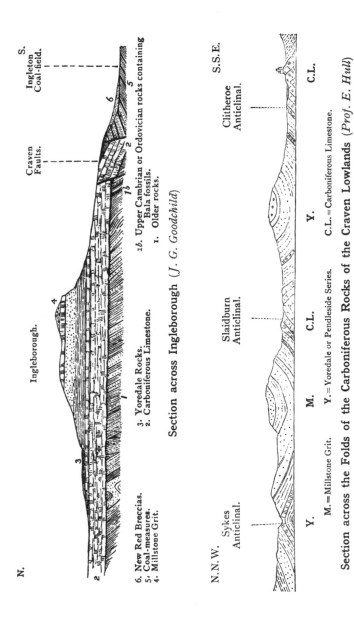

N.

Ingleborough.

Craven
Faults.

S.
Ingleton
Coal-field.

1b. Upper Cambrian or Ordovician rocks containing
Bala fossils.
1. Older rocks.

3. Voredale Rocks.
2. Carboniferous Limestone.

6. New Red Breccias.
5. Coal-measures.
4. Millstone Grit.

Section across Ingleborough (*J. G. Goodchild*)

N.N.W.

Sykes
Anticlinal.

Slaidburn
Anticlinal.

Clitheroe
Anticlinal.

S.S.E.

Y.

M.

C.L.

Y.

C.L.

M. = Millstone Grit.   Y. = Voredale or Pendleside Series.   C.L. = Carboniferous Limestone.

Section across the Folds of the Carboniferous Rocks of the Craven Lowlands (*Prof. E. Hull*) north, and those south of the faults, is that instead of being almost horizontal, the latter, owing to lateral compression at the close of Carboniferous times, have been thrown into a series of folds, each fold being from one to several miles across. The axes of these folds run from E.N.E. to W.S.W., therefore the pressure which produced them must have been at right angles to that direction.

Leaving the area of the Craven Lowlands, we may say something more about the Millstone Grit. It occupies a very large area in the riding, extending from Great Whernside southwards in a band, 16 to 26 miles wide, to Shipley, and thence, often 10 miles broad, to the neighbourhood of Sheffield. It is from 2500 to 3300 feet thick, and forms the main mass of the Pennine Chain, the structure of which, south of Haworth to the southern boundary of the riding, consists of a broad anticlinal arch of Millstone Grit from which the strata dip westwards to the Burnley and South Lancashire coalfield, and eastwards to the Yorkshire coalfield. The grit is sometimes worn by wind and weather into the most fantastic forms, as at Brimham Rocks near Pateley Bridge, and Plumpton Rocks near Harrogate. The soil it yields is generally poor.

We now come to the Coal Measures, which, from an economic point of view, are of far more importance than any other rocks. They agree with the Millstone Grit in consisting almost exclusively of shales, sandstones, coals, and underclays. Their chief fossils are remains of land plants, then shells of freshwater molluscs, and

c

lastly, sea-shells in a few thin marine bands.   The Coal
Measures occur in a synclinal or basin fold of which the

The Idol Rock, Brimham Rocks (*Copyright by Mr G Bingley*)

axis runs from N.N.W. to S.S.E.   Only the western
portion of this basin is exposed, the eastern portion

being concealed under Permian and Triassic rocks. The exposed portion of the basin extends from near Leeds to near Nottingham. In the West Riding, the northern boundary runs from west to east a few miles north of Bradford and Leeds, and here the Millstone Grit

Plumpton Rocks (*Copyright by Mr G. Bingley*)

(*Showing Weathering of Millstone Grit*)

dips southward under the Coal Measures. The outcrop or exposed portion is here 23 miles broad. The western boundary from Denholme near Haworth to Sheffield is formed by the Millstone Grit dipping eastward under the Coal Measures, and the outcrop is 13 miles wide at Sheffield. The eastern boundary is formed by the overlying Permian Magnesian Limestone. The length of Coal Measure outcrop from north to south within the riding is 34 miles.

The Coal Measures have been divided into Lower, Middle, and Upper. The Lower Measures, from the top of the Millstone Grit to the base of the Silkstone Coal, are from 900 to 1600 feet thick, and contain few coal seams of importance. Among them the Gannister or Halifax Hard Coal, from 1 to 4 feet thick, is remarkable in three respects. Its " seatstone," or so-called under-clay, consists of a very hard rock, composed chiefly of quartz with rootlets, which is valuable for lining steel melting furnaces. Its roof contains marine fossils, and the coal itself contains the so-called " coal-balls," calcareous nodules in which are excellently preserved fossil plants. The Middle Coal Measures, from the base of the Silkstone Coal to the base of the Etruria Marl Group, are from 1800 to 3000 feet thick, and contain by far the most valuable seams. (See p. 80.) The only exposure of Upper Coal Measures in the riding is at Conisborough, where a few feet of shale and sandstone underlie the Permian rocks. A small coal-field occurs west and north-west of Ingleton.

The conditions under which the coal seams were formed were probably as follows. Large areas of low-lying ground, little above or even below sea-level, became lakes, swamps, or estuaries, the former separated from the sea by barriers of no great height. In these swampy areas a dense vegetation flourished. It con-sisted largely of ferns, Pteridospermeæ (fern-like plants bearing true seeds), plants such as Calamites, allied to our horsetails (Equisetum), and Sigillaria and Lepidodendron, allied to our club mosses (Lycopodium) but forming forest trees, and accompanied by many other plants. These continued to grow for long periods, and their remains were no doubt mixed with those of other vegetation drifted into the area. Then subsidence took place, and the vegetable matter was covered by beds of sand and mud, until the swamp, lake, or estuary was sufficiently silted up for vegetation from higher ground again to spread over it, and ultimately form a second coal bed, separated from the first by beds of sandstone and shale. Subsidence, with intervals of rest, would account for the great thickness of the Coal Measures as a whole. Occasionally the sea burst the barriers or flowed in as the result of the subsidence, as we have seen in the case of the Gannister coal.

At the close of the Carboniferous period, great mountain-making forces came into play, and lateral pressure caused the originally horizontal strata to be folded so as to produce the great anticlinal arch separating the Yorkshire from the Lancashire coalfield. This fold is accompanied by a fault termed the Anticlinal Fault, which runs by the Saddleworth valley to Todmorden and the western slopes of Black Hambledon. Pressure, acting from S.S.E. to N.N.W., ridged up the rocks of the Craven Lowlands. Great fractures of the strata, accompanied by vertical displacement, either originated at this time, or increased greatly. The chief of these were the huge Dent Fault, running S.S.W. from near

Kirkby Stephen to Leck Fell, by which Carboniferous
rocks on the downthrow side have been jammed from the
east against the Silurian ; and secondly, the Craven
faults, three in number, of which the northern runs
E.S.E. from Leck Fell to near Pateley Bridge, with a
downthrow to the south of over 600 feet at Jenkin Beck.
It throws down the Carboniferous Limestone against
the Ordovician and Silurian of the inliers.   An almost
parallel fault, the Middle (often called the South) Craven
fault, throws down the Ingleton Coal Measures and
Permian, and farther eastward the Yoredales and Millstone Grit on the south against the Carboniferous Limestone.   The result of the Middle Craven Fault has been
to give rise to Giggleswick Scar, Malham Cove, and
Gordale Scar.   It has been calculated that the vertical
displacement of the strata (for instance the Millstone
Grit of the summit of Ingleborough) brought about
by combined faulting and folding, amounts to not less
than 5375 feet near Ingleton.   A third, and southern
fault runs south-eastward from near Settle to Gargrave
with downthrow to south-west.

The result of these folds and faults was that the
Carboniferous rocks were elevated considerably above
sea-level, and erosion by rain and rivers set in, and
continued for a very long period, so that north of Leeds
a thickness of at least 3000 feet of Coal Measures was
worn away before the next succeeding Permian rocks
were deposited.   At this time, too, the erosion of the
Coal Measures at the summit of the anticline of the
Pennine Chain began, which has ended in completely

separating the once continuous Yorkshire and Lancashire coalfields.

At last the Carboniferous and older rocks sank sufficiently to form two great enclosed sea basins, comparable to the present Caspian Sea, and most probably separated by the anticlinal arch of the Pennine Chain. The Permian deposits of these two basins differ so much that they form two distinct types (see table), an eastern type from 430 to 570 feet thick, eastward of the Pennine Chain, and a western type west of it, both probably deposited under desert conditions. In the eastern basin, which extended from near Nottingham into Northumberland, the most important deposits are those of Magnesian Limestone. The Permian rocks overlie the Carboniferous with complete unconformity, for at Conisborough they succeed the Upper Coal Measures, whilst N.E. of Leeds they lie on Millstone Grit. The eastern type of Permian has an outcrop from $2\frac{1}{2}$ to 8 miles broad, and dips gently eastwards under the overlying Trias. It usually forms an escarpment facing west, and the limestone gives rise to a good, light soil. Beautiful gorges occur where the rivers cut through the Magnesian Limestone. The western type of Permian strata occurs within the riding in only three small outliers, the largest 2 miles west of Ingleton, the second in the banks of the Ribble, west of Clitheroe, the third at Bashall Brook, 3 miles N.W. of Clitheroe. The first consists of red sandstones and conglomerates overlying the Coal Measures, the second consists of red sandstone and marl, and is remarkable because it overlies Carboniferous

Shales with Limestone, a fact which is believed to indicate that 15,000 feet of newer Carboniferous strata were worn away before it was deposited. The third is composed of bright red false-bedded sand.

Of the Triassic rocks the Bunter usually consists of

Magnesian Limestone overlying Millstone Grit Banks of the River Nidd at Knaresborough

(*Copyright by Mr G. Bingley*)

reddish sandstone ; the Keuper of sandstone below, and red marl with gypsum above. The sandstones give rise to a light sandy soil, the marl to a stiff red clay. Both divisions are, however, very variable in Yorkshire, and are seldom exposed, being usually covered by glacial drift (boulder clay, sand, and gravel) or other superficial deposits. They were probably deposited in salt lakes by wind and rivers under desert conditions. The Trias overlies the Permian rocks, and extends under the Vale of York. There is no doubt that the outcrop of the Permian and Triassic rocks of the West Riding originally extended much farther west than it does now, and Dr J. E. Marr and the late Mr A. J. Jukes-Browne have expressed their belief that the Keuper and overlying Rhætic, Liassic, Oolitic, and Cretaceous strata, now exposed in the East Riding, once extended over the anticlinal of the Pennine Chain, and that the final elevation of the chain took place in Tertiary times. Be that as it may, no trace of them remains there, and the deposits next met with in the West Riding are not only newer than the Cretaceous, but later than all the Tertiary rocks, being of Pleistocene age. Between the Triassic and the Pleistocene a vast period of time elapsed, and the Pleistocene deposits are necessarily unconformable to the older strata everywhere in the West Riding. Space will not permit more than a glance at the Pleistocene glacial deposits, which consist of boulder clay (believed to be the bottom moraine of glaciers), sand, and gravel, and ice-transported erratic blocks. A great glacier from the southern uplands of Scotland and the Lake District crossed the Pennine Chain by the Stainmoor Pass, N.E. of Kirkby Stephen, and turned southeastward down the Vale of York, where it has left two concentric crescent-shaped terminal moraines, the outer one running through Escrick, the inner one through York. Local glaciers descended Wensleydale, Nidderdale, Wharfedale, and Airedale, and have left traces in the form of lateral or terminal moraines. The Calder valley does not appear to have had a glacier of its own, but a lobe of the Ribblesdale glacier and of a great north-western glacier reached its upper end, and the waters of

Norber Boulder (*Copyright by Mr G. Bingley*)
(*Erratic block of grit resting on Carboniferous Limestone*)

ice-dammed lakes escaped down its gorge. There is uncertainty about the valley of the Don, but at Balby, near Doncaster, there is one of the finest exposures of boulder-clay (containing Shap granite) in the county. Ribblesdale supported a great glacier, and in the neigh-bouring Crummackdale erratic blocks of Silurian grit have been transported uphill by ice, and left stranded on a Carboniferous limestone pavement at Norber Brow.

River terraces serve to indicate the various levels at which rivers formerly flowed, the highest terrace being, of course, the oldest ; they occui in many places, *e.g.* the junction of the Aire and Calder, and in Wharfedale between Ilkley and Arthington.   Alluvium—the mud and other material deposited by rivers where their current is insufficient to transport it—occurs along all the larger streams and rivers, and especially in the Vale of York. Peat covers many square miles of the Pennine moors, especially where cotton-grass (*Eriophorum*), and to a less extent where ling (*Calluna*) predominates.   Lowland peat occurs at Goole Moor.

7. Caverns, "Pots," Underground Drainage, and Fossil Mammalia

We have already alluded to many instances of underground drainage.   Another noteworthy example is that of Dale Beck, which, under the name of Little Dale Beck, rises between Whernside and Blea Moor.   It flows underground in many places under normal conditions, but when in flood occupies a well-worn surface channel. West of Ribbl head Station it is swallowed up, but reappears after some hundreds of yards from two neighbouring openings, at the lower end of Gatekirk Cave.   It passes underground again, and emerges a mile lower at Weathercote Cave, where it forms a beautiful waterfall

75 feet high.   A hundred yards beyond is a fissure, 80 feet long and 50 feet deep, called Jingle Pot, usually dry; and 200 yards further, a more circular cavity, 30 feet deep and 90 feet across, called Hurtle Pot, which is connected with the underground channel.   Nearly a mile below Weathercote, at God's Bridge, the stream finally emerges from beneath the beds of limestone.

The " pots," sometimes called pot-holes, are met with in limestone only, and are due partly to the chemical, partly to the mechanical, action of water and its burden of gravel acting upon joints or fissures.   The pots in the limestone of Ingleborough are, as might be expected, common, numbering nearly fifty.   Alum (Helln) Pot is a huge chasm on the eastern flank of Ingleborough, 130 feet long and 290 feet deep.   A small stream precipitates itself over its lip, and the waters of Alum Pot Beck pass through Long Churn cave, and find their way into the pot by a waterfall through the roof.   The water of Alum Pot passes underground beneath the River Ribble, at the bottom of the valley, and rises up on the opposite side of the valley in a spring, called Turn Dub.

Gaping Gill, a well-known pot on the south flank of Ingleborough, 2 miles north of Clapham, is 365 feet deep, and has an elliptical opening only 30 by 15 feet in diameter.   At the bottom is an enormous chamber, 479 feet in length and 110 feet in height, from which passages and chambers extend for an explored distance of nearly two miles.   Into this chasm the Fell Beck precipitates itself to reappear a mile away, at Clapham Beck Head, close to the entrance of Ingleborough Cave.

Gaping Gill Hole

Ingleborough Cave is in fact part of the original subterranean course of Fell (Clapham) Beck, and is still
used in time of flood.   The cave has been explored for
900 yards, and is celebrated for its stalactites, as are
Yordas Cave in Kingsdale, 4 miles north of Ingleton,
and Stump Cross Caverns, 5 miles west of Pateley
Bridge.

Several caves of the West Riding have a special
interest, on account of the remains they contain of bones
of animals which existed in England not only during
the later Prehistoric period, but also in the Pleistocene.
Traces of early man, too, are in some cases found in them.
The best known and most interesting is Victoria Cave,
2 miles north-east of Settle, in which the following
succession of deposits was found :—

8. Talus of angular blocks of limestone, 2 feet.

7. Romano-Celtic Layer (bones, fragments of pottery,
charcoal, and burnt stones), 2 feet.

6. Talus of angular blocks of limestone, 6 feet.

5. Neolithic (see Chap. 10) Layer, resembling No. 7
except in its contents.

4. Talus of limestone blocks, maximum thickness 19
feet.

3. Upper Cave Earth, with bones of existing species
of mammals.

2. Laminated Clay.

1. Lower Cave Earth, with bones of Pleistocene
mammals.

This last-named stratum yielded remains of weasel,

marten, hyæna, *fox, *grisly bear, *brown bear, bison, *red deer, goat, *reindeer, *Rhinoceros leptorhinus, Hippopotamus, Elephas antiquus,* and *Bos primigenius* (Urus). The whole of this group is of Pleistocene age, and before the glacial period, as is shown by the fact that the Lower Cave Earth is overlain by ice-borne boulders at the entrance of the cavern.

In the Upper Cave Earth the animals marked with an asterisk are found, as well as remains of badger, horse, pig, and sheep, and traces of man are shown by bones cut with some sharp instrument. The Neolithic layer contained a bone harpoon of Mas d'Azil type, a bone bead, three rude flint flakes, and broken bones of brown bear, stag, horse, and Celtic short-horned ox (*Bos longifrons*). From the thickness of intervening talus it has been estimated that the Neolithic layer may be 3600 years older than the Romano-Celtic. In the latter layer were found bronze ornaments, brooches (some of them enamelled), bracelets, a finger ring, a buckle, bone pins, studs, Roman bronze and silver coins dating from 117 to A.D. 353, and barbarous imitations of these of about A.D. 400–500. These relics were probably left by Romanised Britons, fleeing before the Saxon invasion in the fifth, sixth, or early seventh century.

Another bone cave, with similar Romano-Celtic remains, is Dowkerbottom Cave, in the valley of the Skirfare, 1½ miles north-west of Kilnsey. The bones found were all those of existing species. In Elbolton Cave, near Thorpe, south of Grassington, Neolithic pottery was discovered, associated with bones of

mammals of existing species.   In Raygill Fissure in the
Lothersdale anticlinal, 5 miles south-west of Skipton,
remains of elephant, *Rhinoceros leptorhinus*, hippopotamus, teeth of lion, etc., of Pleistocene age, have
been found.

8. Natural History

Great Britain and Ireland are continental islands of
geologically recent origin ; that is, their separation from
the Continent is recent.   That the British Isles have
formed part of continental Europe is shown not only
by the fact that remains of the Pliocene and Pleistocene
species of mammals of the British Isles occur as fossils
on the Continent, but also by the fact that the existing
animals and plants of these islands closely agree with
those of the Continent.   On the other hand, if the separation had existed for a long period many peculiar British
forms would certainly have been evolved.   The separation is believed to have taken place at the Straits of
Dover in late Pleistocene times, and Ireland was probably separated from England rather earlier ; as a result
of which Ireland has only four species of reptiles and
amphibia, Britain has thirteen, and Belgium twenty-two
species.   Owing to the limited area, the restricted range
of climatic and other conditions in the British Isles, and
no doubt also to the short time which elapsed after the
glacial epoch before separation from the mainland
occurred, they are less rich in species of both animals
and plants than the European continent.   The vast

majority of the British animals and plants have been derived from the Continent, and the few forms peculiar to Britain are probably due to changes brought about by isolation.

The distribution of vegetation is chiefly determined by soil conditions and climatic conditions. The former depend primarily on the geology, but in cultivated areas are modified by human agency ; the climatic conditions depend mainly on altitude, and the temperature and rainfall which, in their turn, are closely connected with it.

The vegetation of the West Riding may be grouped under three heads, viz. that of (1) farmland, (2) woodland, (3) moorland. It has also been arranged according to zones of altitude, viz.—

1. Zone of wheat cultivation and lowland oak wood : upper limit 500 to 700 feet.

2. Upland oak wood and greater part of cultivation without wheat : upper limit 900 to 1000 feet.

3. Birch wood and pine wood : upper limit 1250 feet, which is also the lower limit of the cotton-grass moss.

4. Heather moor, grass heath, and limestone pasture : upper limit 1500 to 1600 feet.

5. Cotton-grass moss and bilberry summit : upper limit 1800 to 2000 feet.

6. Alpine pasture, 2000 to 2300 feet.

The area in the West Riding which can be described as Alpine pasture is very small, and occurs only on the summits of some of the highest hills, such as Ingleborough

D

and Whernside, where one or two arctic willows grow. The purple arctic saxifrage (*Saxifraga oppositifolia*) is a feature of the crags of Upper Scar Limestone, just below the summit plateau of Ingleborough and Penyghent.

The cotton-grass moor (or *Eriophorum* moor) occupies a larger area than any other natural plant association of the Pennine Chain. These moors or " mosses," as they are more commonly called, are largely composed of closely-set tussocks of the cotton-grass, more accurately the cotton-sedge (*Eriophorum vaginatum*). For the greater part of the year these moors are drab and dismal to an extreme degree, but in early summer the *Eriophorum* bears dense white cottony fruiting heads, which somewhat relieve the monotony. The cotton-grass in its decay—either alone or together with some of its associates, such as the various bog-mosses, ling, the cross-leaved bell heather, and the crowberry—forms abundant peat, which may accumulate to a thickness of from 5 to 30 feet. The cotton-grass moor occurs at heights of over 1250 feet, chiefly on the shales and grits of the Millstone Grit series and Yoredale beds, and its spongy mass affords abundant supplies of water to the reservoirs near the heads of the valleys. Cotton-grass moor caps Baugh Fell, Widdale Fell, Blea Moor, Fountains Fell, and Darnbrook Fell, part of Whernside, and Penyghent, and occupies vast stretches in many localities, notably from Stoodley Pike, near Todmorden, to Margery Hill, near the Derwent, a distance of 22 miles, and continuing as mixed cotton-grass and heather moor to Doveston Tor, a further 4 miles.

A variant of the cotton-grass moor occurs locally where the bilberry (*Vaccinium Myrtillus*) is equally dominant with, or largely displaces, the cotton-grass. This occurs where the peat has been much intersected and consequently well drained by a system of runnels. Bilberry forms another distinct association on dry, rocky, wind-swept screes and summit-ridges, such as Earl Seat and Simon Seat, north of Bolton Woods, West Nab near Meltham, and Holme Moss (Black Hill).

Heather moor, so important for grouse rearing, is that in which heather or ling (*Calluna Erica*) predominates, associated with bilberry, cowberry, rushes, etc. The heather grows best in well-drained places on peat, on the lower watersheds or the lower slopes of the higher hills, immediately below the cotton-grass moors. It seldom occurs on limestone.

Grass heath consists of natural grass, with heath plants subordinate, and is of two types—(1) The wet grass heath, which is flat or nearly so, somewhat peaty, and dominated by the blue moor grass (*Molinia*); and (2) the dry grass heath, which occurs on steep slopes, is almost without peat, and dominated by mat grass (*Nardus stricta*). Grass heath usually forms a fringe round the moorland, and may occur on limestone when the latter is covered by a layer of glacial clay or humus. The true limestone pasture consists of the short, sweet, and nutritious grass growing on the thin soil of the Carboniferous limestone and the Permian Magnesian Limestone.

The natural woods of the area are not extensive. They are chiefly oak woods on the sandstones and shales, ash woods or copses on the limestone slopes of the Craven dales, and beech woods on the Magnesian Limestone. Birch woods or thickets occur above the level of the upland oak woods, and formerly extended up to 1725 feet, as is shown by stems in the peat. No natural pine woods now exist, though Scots pine stems are found in the peat on the moorlands, showing that pine woods formerly occurred. The area, in 1913, of woods, plantations, and coppice in the riding, was 67,629 acres.

Some thirty species of mammals inhabit the West Riding, among which are at least eight species of bats. The pine-marten and the polecat are probably extinct or nearly so. Both the badger and the otter are common ; the other species call for no special remark, unless it be the blue or mountain hare, which is not a native, but has been introduced in various places, with no very marked success.

The great range of variation in the physical character of the country is naturally favourable to a rich avifauna, which is what we find. The number of species of birds which have been recorded as occurring in Yorkshire as a whole is 326, and the great majority of these are no doubt found in the West Riding. A rather remarkable number of rarities have occurred, among them the little bustard, White's thrush, the shore lark, cream-coloured courser, blue throat (*Cyanecula suecica*), and others. The waxwing has occurred in considerable numbers in some winters, and the crossbill is thought to build regularly. Contrary to usual belief, the nightingale is not uncommon, and it nests in many parts. It is probably extending its range farther north, and increasing, as the starling is, to an unwelcome extent. The hawfinch is also increasing in numbers. But many species show a regrettable diminution in numbers. The peregrine is now a rarity, and as a breeding species is described as being " almost doomed." Of the raven much the same might be said. The Norfolk plover, or thicknee, breeds, but only in one or two localities. The Yorkshire grouse moors are among the best in the kingdom and the Blubberhouses Moors of the West Riding are especially celebrated. The Yorkshire birds are heavier than the Scotch.

9. Climate

Climate denotes the average atmospheric conditions of any region. The climate of Britain is what is known as marine or insular, the effect of the nearness of the sea, owing to the great specific heat of water, being to moderate both the heat of summer and the cold of winter, and render the climate more equable than that of continental areas. The drift of warm water from the south-west of the North Atlantic, and the prevalence of south-west winds increase the effect due to the presence of the sea. Although the West Riding is an inland area, these factors have nevertheless a great influence on its climate.

From a climatic point of view the riding may be roughly divided into two chief regions, that of the Pennine Chain on the west, with a cool and wet climate, and that of the Coalfield Lowlands and Vale of York on the east, of a warmer and drier type, but there are, of course, many places with a climate intermediate between the two. Temperature is one of the most important features of climate. It depends on latitude, altitude, distance from and temperature of the sea, and exposure. In the West Riding differences of temperature due to latitude are small. If we examine a map of Britain showing isotherms (that is lines drawn through places having the same temperature, the observed temperatures being reduced to their value at sea-level), we find that in January the isotherms in Yorkshire run almost north and south, the west, at corresponding levels, being rather over 1° F. warmer than the east, so that the effect of latitude is not apparent. In July the isotherms run nearly east and west, but the mean shade temperature at sea-level is 61° in the south, and 60° in the north of the riding, a difference of only 1°. The effect of altitude is 800 times as great as that of latitude, that is, temperature diminishes vertically far more rapidly than horizontally, measuring towards the pole. This vertical temperature gradient is variable according to the month and locality. The Meteorological Office assumes 1° F. for 210 feet for the annual mean of maximum thermometer readings, 1° F. for 250 feet for minimum readings. The mean of these two is 1° F. for 230 feet. On this assumption the summit of Whernside would have an average temperature 8° F. lower than that of Halifax.

TABLE OF APPROXIMATE MEAN TEMPERATURE,
- RAINFALL, AND SUNSHINE

| Locality. | Altitude. Feet. | Temperature. | | | | Rainfall. | Rain Days. | Driest Month. | Sunshine Hours. |
		Jan.	July.	Range.	Annual.				
York	56	37.7	60.4	22.7	48.1	24.8	188	Feb.	1273
Bawtry . . .	65	37.2	60.5	23.3	48.0	23.4	166	Feb.	..
Wakefield . .	96	38.2	59.9	21.7	48.0	26.2	168	Feb.	..
Leeds	131	39.5	61.9	22.4	49.6	24.7	..	Sep.	1237
Bradford . .	366	36.5	59.6	23.1	47.2	30.9	..	Feb.	..
Huddersfield . .	410	36.9	59.3	22.4	47.1	34.3	..	April	1167
Sheffield (Weston Park) .	429	38.2	60.4	22.2	48.4	29.9	182	April	1315
Harrogate . . .	479	36.6	58.4	21.8	46.6	29.6	197	April	1442
Giggleswick . . .	500	36.9	57.6	20.7	46.2	43.4	..	{ April and May	..
Halifax . . .	530	35.9	56.8	20.9	45.6	37.2	184	May	..

The wettest month, on the average, is in each case October, except at Giggleswick, where it is December, but the wettest and driest months vary in individual years.

Although no place mentioned in the preceding table lies at a great elevation, the figures give a good idea of the climate of the zone in which most of the population is concentrated. Halifax at 530 feet is decidedly cooler, both in summer and winter, than York at 56 feet. On the other hand, York and Bawtry, low-lying in the plain, have a greater range of temperature (York, 22.7° ; Bawtry, 23.3°) between January and July than Halifax (20.9°), showing that their climate is slightly less marine or more continental. As the rainfall map shows, there is a great contrast between the highlands of the Pennines and the low ground east of them.   The winds from the west and south-west, laden with invisible aqueous vapour from the Atlantic and Irish Sea, on reaching the western side of the Pennines, are compelled to ascend, and as they ascend, the air becomes cooled, and by a well-known law precipitates the aqueous vapour it contains as rain, snow, or hail—hence a rainfall of from 40 to 70 inches on the Pennine summits.   At Whitendale (830 feet) 4 miles N.W. of Slaidburn, the average rainfall is 73 inches.   At Oughtershaw Hall, at 1175 feet, near the source of the Wharfe, it is 71 inches. The Pennine areas of great rainfall are valuable as gathering-grounds for rivers and reservoirs, and the absence of population prevents pollution.   A remarkable indentation in the isohyets (lines of equal rainfall) occurs along the valley of the Aire, Skipton, at 350 feet, having only 34 inches of rain.   On the eastern side of the Pennine Chain condensation is perforce lessened, most of the aqueous vapour having been already parted with.   Although Dunford Bridge, at 954 feet, has 49 inches of rain annually, the rainfall at Sheffield, 29.9, and Barnsley (350 feet), 27 inches, is moderate, and in the plain east of the hills, where there are no considerable elevations to condense vapour, the fall does not exceed 25 inches, as at York and Bawtry, or at Doncaster, where at 32 feet it is about 23 inches.

In hours of bright sunshine the West Riding stands considerably below the south and south-east coast of

England; *e.g.* St Leonards (Hastings) has 1795 hours, and Clacton even more. How much smoke diminishes sunshine is shown by comparing Weston Park, Sheffield (average for 1909–14), 1286.9 hours, with Attercliffe, in the smokiest part of Sheffield, 978 hours.

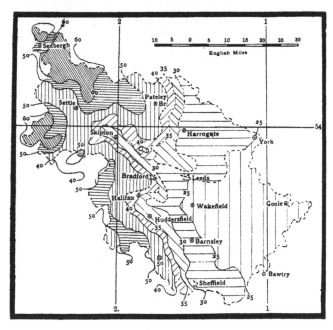

Average Annual Rainfall of the West Riding
By Dr H. R. Mill, of the British Rainfall Organization

10. People—Race, Language, Settlements, Population

History affords us scant information as to the inhabitants of Britain before the Roman invasion in 55 B.C., but by the aid of careful exploration of caves and other sites much has been learnt. The bones of man and domestic animals, implements, and ornaments, found in caves or river-deposits, throw considerable light on the subject. Four main stages of culture and corresponding Ages have been recognised—the Palæolithic or Old Stone Age, the Neolithic or New Stone Age, the Bronze Age, and the Iron Age. It must not be supposed that implements of the older type were immediately abandoned when the newer type was introduced. Stone implements, for instance, continued to be used to a certain extent in the Bronze Age : they are, indeed, in use to this day among many savage tribes. Palæolithic man fashioned his implements by chipping flint, chert, or quartzite into spear-heads, arrow-heads, scrapers, borers, and saws. With flint tools he cut bones and antlers into spear-heads, harpoons, and needles, the latter being used to sew skins together with reindeer sinew for clothing. He lived by hunting and fishing. His food was either roasted, or boiled in water, into which he dropped hot stones, for he probably used vessels made of skin or wood, as he appears to have been unacquainted with the art of making pottery until towards the close of the Palæolithic Age. No undoubted remains of Palæolithic man have been found in the West Riding, though in two caves at Creswell Crags, in Derbyshire, only three miles from the boundary of the riding, more than a thousand implements of this Age were discovered.

Palæoliths are only chipped, but Neolithic man ground and polished such of his implements as would be thereby improved, though he used others, such as arrow-heads, merely chipped. The axe was his most important tool, but he also made knives, scrapers, saws, etc. He inhabited caves and partly sunk dwellings, the remains of which we see to-day as " hut circles." He had a knowledge of spinning and weaving flax, and probably wool, and he made pottery shaped by hand, not turned on a wheel. He cultivated wheat, and possessed domestic animals such as the dog, sheep, goat, short-horned ox, horse, and pig. He made canoes from tree trunks, and buried his dead usually in long, oval barrows, but sometimes in round barrows. In Yorkshire the bodies were burnt before burial. He left behind him various stone monuments, though these are scarce in the West Riding, the best example being the well-known Devil's Arrows at Boroughbridge, menhirs or standing stones, believed to be sepulchral. The Neolithic relics of Victoria Cave and Elbolton Cave have been previously mentioned, but Neolithic implements have occurred in most of the parishes in Yorkshire. From numerous skeletons, the average stature of Neolithic man has been calculated as 5 feet 5 inches. His skull was long and narrow, the ratio of length to breadth (the cephalic index) being about 100 : 70. It is generally held that he was of old Iberian race, of which the Basques of the Pyrenees may be considered a remnant, but his descendants survive in Wales, in Scotland among the small dark Highlanders, and in south-west Ireland.

In Britain the transition from the use of stone to that of bronze appears to have been brought about between 2500 and 1400 B.C. by the invasion of people armed with bronze weapons. The earlier bronze axes resembled stone axes in form. Later, flanged axes, and still later, axes with lateral stop ridges or palstaves and socketed celts were made. The men of the Bronze Age had bronze daggers, and later, bronze swords and spears and javelin heads, though they also used flint arrow-heads. They decorated themselves with ornaments of bronze, amber, and gold, and dwelt in some cases in round stone huts, and also in lake-dwellings. Their skulls were broad, the cephalic index being about 81, and in height they averaged 5 feet 9 inches. Probably six or seven centuries before the Christian era Britain was invaded by the Goidels or Gaels. Their language was Gaelic, now a dialect of Ireland, Scotland, and the Isle of Man.

Probably about 350 B.C. the Brythons, from whom the name of Britain is derived, invaded the island and brought with them the knowledge of the working of iron, though the metal was no doubt introduced much earlier. Their language was later represented by Welsh, Breton, and Cornish. In the early Iron Age, bronze continued in use for sword hilts and scabbards, shields, the metal

parts of harness, and ornaments ; but swords were made
of iron, and arrows, daggers, spears, and javelins were
tipped with that metal.   Chariots had iron tyres.   The
average height of the men and women of this early Iron
Age ranged from 4 feet 6 inches to 5 feet 7 inches, and
averaged only 5 feet 1 inch.   Their skulls were long,
the cephalic index averaging 72.4.

The Roman conquest and dominion seem to have had
little effect on the British races.   It was not until 627
that the Angles, under Edwin, completed the conquest
of the British kingdom of Elmet, corresponding approximately to the West Riding.   The stature of the Anglo-
Saxon men and women of East Yorkshire ranged from
4 feet 10 inches to 6 feet 6 inches, and averaged 5 feet
4 inches, and their skulls were long, the average cephalic
index being 72.4.   Their language, commonly spoken of
as Anglo-Saxon, is Old English.   Characteristic Anglian
terminations of place names are—*ley*, a meadow ; *bridge*,
croft ; *den*, a valley or swine-pasture in a wood ; *ham*,
a home or enclosure ; *hill, wood* ; *worth*, a holding.

In 867 the Danes took York, and in 876 the Kingdom
of Deira was divided among them.   Their invasion and
conquest had more ethnological result in Yorkshire than
in most other parts of England.   Between the ninth and
eleventh century the Danes arrived in considerable
numbers.   Their settlements are indicated by the ter-
mination *by*, meaning a village or town, as in Kirkby,
Maltby, Selby, Sowerby, Wetherby ; and *thorpe*, a
village, as Cowthorpe, Thorpe Salvin, etc.

In 1801 the population of the ancient or geographical

county of Yorkshire was 859,133 ; in 1911, it was 3,954,844. The West Riding (registration area) had in 1801 a population of 574,681 ; in 1911, 3,044,608, but with York 3,127,659. The number per square mile in 1911 was 1099 exclusive, and 1126 inclusive, of York. The figures per square mile are rather misleading, for while the summits of the Pennine Chain are uninhabited, and the Vale of York has only a moderate density of population, the coalfield is densely populated. More than half the population of the riding is concentrated in Sheffield, Leeds, Bradford, Huddersfield, Halifax, Rotherham, Dewsbury, Wakefield, and Barnsley.

11. Agriculture

Although better known as the workshop of England, the West Riding is really a great agricultural county. The number of persons employed in agriculture in the administrative county (with the county and municipal boroughs and the city of York), in 1911, was 52,040, of whom 40,161 men and 4386 women were farmers, graziers, and farm-workers ; 6319 men and 174 women were gardeners (not domestic), nurserymen, seedsmen, and florists, and the remaining 1000 were woodmen, agricultural machine owners or attendants, etc. The number of agricultural holdings in the riding in 1914 was 25,929, twice as many as in the North Riding, and more than three times those in the East Riding. Of this total, 19,455 were above 1 and not exceeding 50 acres, 4613 above 50 and not exceeding 150 acres, and 1861 above 150 acres, the total acreage of the holdings being 1,168,154, and that of the average holding, 45 acres. In the East Riding the average holding is 91 acres, in the North Riding, 67 acres. The West Riding is thus a county of small holdings.

Of the 1,168,154 acres, 341,755 acres are arable land, and 826,399 acres permanent grass (the latter not including 232,187 acres of mountain and heath land, used for grazing), the proportion of arable to permanent grass being 29 per cent. arable to 71 per cent. pasture. In the East Riding the proportions are reversed, 66 per cent. arable, to 34 per cent. pasture ; in the North Riding, 37 per cent. arable, to 63 per cent. pasture. It is thus evident that the West Riding is much more a pastoral or stock-breeding, than a corn or crop-growing county. Nevertheless the area of crops is large. As to wheat, the area is 50,522 acres, which compares with 69,145 in the East Riding, and 28,425 in the North Riding. Nine counties of England have a larger area of wheat. In 1867 the area under wheat in the West Riding was 97,034 acres. The diminution in area is (speaking of pre-war conditions), due to the fall in the price of wheat. In the seven years 1861–7, the average price was 50s. 3d. per quarter ; in 1908–14, it was 33s. 4d. In the West Riding, wheat is grown up to a level of 600 or 700 feet, but in the narrow western valleys, only up to 400 feet. The chief area of wheat is in the low ground east of the Pennine Chain. The area of barley or bere in the riding is 49,754 acres, almost as large as that of wheat. In 1867 it was 73,123 acres. The average price of barley,

Liquorice in cultivation near Pontefract

1861–7, was 34s. 7d., but in 1908–14, only 26s. 10d. Oats occupy 74,407 acres, compared with 65,357 acres in 1867. Oats and barley can be grown up to a level of 1000 feet. Compared with these areas, the space devoted to rye (3405 acres), beans (1250 acres), peas (5112 acres), seems small. Potatoes occupy 25,815 acres—more than twice the area in either the North or East Riding. Very little small fruit—639 acres—is cultivated, but no less than 2980 acres of rhubarb, which is far more than is grown in any other county. A crop for a great number of years cultivated to a small extent near Pontefract is liquorice, a papilionaceous plant, with roots 3 or 4 feet long, the juice of which is made into stick or " Spanish " liquorice and " Pomfret cakes," and is also used in medicine.

As regards live stock, the number of horses used for agricultural purposes was 35,443, and of all other horses, 27,571, total 63,014, and it is worthy of note that this is about 50 per cent. more than in either the East or North Ridings. The number of cattle, too (272,594), was nearly three times as many as in the East Riding, and one and a half times as many as in the North Riding.[1]

---

[1] The above remarks are based on the 1914 statistics as being the last normal ones. In 1918 flax was grown on 1994 acres.

E

12. Industries—(a) Textile Industries

The arts of spinning and weaving, as already stated, were known to Neolithic man.   Spindle-whorls and bone combs for beating the weft into position on the warp, have been found in his graves ; and in a barrow of Bronze Age at Rilstone, near Skipton, remains of a woollen wrap were found with the body.   Writers of economic history have traced four stages in the development of industry.   These are generally known as the family system, the guild system, the domestic system, and the factory system.   The spinning and weaving of wool were originally carried on at home for the use of the worker's family, but by the twelfth century, weaving had become a commercial craft, regulated by guilds, but carried on by small masters employing two or three men.   In the twelfth and thirteenth centuries England was a great wool-exporting country.   York at that time was the chief town in the county for the weaving of wool.   At the end of the fourteenth century York maintained its pre-eminence in weaving, Ripon and Richmond being next in importance.

In the last quarter of the fifteenth century Halifax, Leeds, Wakefield, Almondbury (near Huddersfield), Barnsley, and Bradford were all weaving cloth.   In the reign of Henry VIII the ancient towns, by oppressive charges and regulations, were driving trade away to the freer western districts.   The guild system was succeeded by the domestic system, which was in vogue from the middle of the fifteenth to the middle of the eighteenth century. The little master lost his position of independence, no longer acted as shopkeeper or merchant, but made his goods for a middleman, who often supplied the raw materials. In the eighteenth century a revolution in the textile industry was brought about by a series of inventions ; firstly that of the "fly shuttle," by John Kay of Bury, in 1733, which doubled the productive power of the weaver ; secondly the "spinning-jenny," invented by James Hargreaves of Standhill, Blackburn, between 1764 and 1767, by which sixteen or more threads could be spun at once ; thirdly the "water-frame," invented or improved by Richard Arkwright in 1769 ; fourthly the "spinning mule," a combination of the methods of Arkwright and Hargreaves, invented by Samuel Crompton in 1779; and lastly the "power-loom," invented by Edmund Cartwright between 1785 and 1792. He also invented two distinct forms of wool-combing machines. These inventions, with the exception of the last two, were at first applied to cotton, and not until a generation later to wool. They necessarily led to the factory system which now prevails.

The table on p. 68 shows the relative importance of the chief textile industries in the riding, exclusive of thread, hosiery, carpet, rug, and felt manufactures.

The West Riding is now the greatest centre of the woollen and worsted trade in the world. While well suited for the pasturage of sheep, its hills were unsuitable for wheat-growing, and its abundant streams furnished the necessary soft water for scouring, as well as the requisite power, before the invention of the steam-engine, now supplied by its coal. It was probably the necessity of supplementing the income derived from sheep-raising that caused its inhabitants to take up woollen and worsted manufacture.[1]

NUMBER OF PERSONS EMPLOYED IN 1911

	Wool and Worsted Manu-facture.	Cotton Manu-facture.	Silk Manu-facture.	Bleaching, Printing, and Dyeing.
Bradford . .	49,036	2,325	3,773	6,761
Dewsbury . .	9,082	481		702
Halifax . .	11,858	2,465	455	1,534
Huddersfield .	16,726	2,034		995
Leeds . .	12,661	769		1,903
Wakefield . .	2,397			140
The rest of the West Riding .	95,820	35,023	3,494	7,261
Totals . .	197,580	43,097	7,722	19,296
Percentage of female em-ployees . .	57%	54%	64%	7.7%

The various branches of the woollen and worsted manufacture are carried on at about 97 localities in the West Riding.   Nearly two-thirds of the wool used is imported. Bradford is the chief centre of the worsted trade, and the headquarters of the great dyers' combination.   In 1800

---

[1] A distinction is drawn between woollen yarns, in which the fibres are crossed, interlaced, and devoid of parallelism, and worsted yarns, in which the fibres are markedly parallelised.

the town had only one mill, in 1850 there were 129, and now over 300. Bradford has a large trade, not only in finished fabrics, but in tops, noils, and yarns.[1] Three miles N.N.W. of Bradford is Saltaire, founded by Mr (afterwards Sir Titus) Salt, in 1850, and renowned

Woollen Mill, Halifax

for its mills producing alpaca and mohair goods, though a dozen other localities make and spin one or both materials. Huddersfield is a great centre for high-class plain and fancy worsted cloths for tailoring. Dewsbury and Batley, as well as Morley, Ossett, and Heckmond-

[1] Tops are balls of combed wool, freed from sand and grease, and ready to be converted into yarn ; noils are the short fibres separated out in the combing process.

wike are centres of the rag wool or "shoddy" trade, but they also produce superior goods. Leeds turns out high-class cloths, as well as large quantities of cloth for the ready-made clothing trade, of which it is one of the

Room in Tapestry Setting Department in a Woollen Mill, Halifax

chief centres. Keighley produces worsted goods in large quantity. Wakefield, in addition to the numbers given above for wool and worsted, employs 200 persons in hosiery manufacture, and the rest of the riding employs 490 (probably more). Blankets are made at Dewsbury, Mirfield, Sowerby Bridge, and a dozen other localities.

Halifax, in addition to producing cloth, employs 1781 persons in the manufacture of carpets, rugs, and felt. Huddersfield employs 570 persons, and the rest of the riding (including Heckmondwike, Liversedge, Brighouse, Dewsbury, and Sowerby Bridge), 2637 persons in the same industry. These places make Brussels, Wilton, tapestry, Axminster, and chenille carpets.

Though the cotton manufacture of the West Riding is less important than that of woollens, it employs over 43,000 persons in 42 localities, and chiefly consists in spinning and " doubling " single cotton yarns into two-fold or threefold threads. The chief localities for spinning and doubling are Barnoldswick, Bradford, Brighouse, Elland, Halifax, Huddersfield, Ripponden, and Sowerby Bridge, but cotton goods are made at Todmorden, Bradford, Earby, Hebden Bridge, Leeds, Skipton, etc.; and sewing cotton at Holmfirth, Leeds, Meltham, and Skipton. The manufacture of silk is carried on at 11 localities in the riding. Most of the mills are engaged in silk spinning. Brighouse is the chief centre, but Low Bentham (near Ingleton), Greetland, and Triangle, near Halifax, and that town itself, and Meltham have spinning mills. Silk fabrics are made at Addingham, Bradford, Leeds, and Ravensthorpe near Dewsbury. Flax and linen manufacture employs 749 persons in Leeds, 477 in Barnsley, and a mill in Knaresborough, and flax and hemp are spun in 9 other localities.

13. Industries—(*b*) Iron, Steel, and other Industries

Whether the Britons were acquainted with any method of reducing the ores of iron is uncertain, but they used the metal in the time of Julius Cæsar. The Romans had extensive ironworks in the Weald and the Forest of Dean, and probably reduced iron ore near Bradford by direct extraction in the malleable state. In 1161 Richard de Busli gave to the monks of Kirkstead, in Lincolnshire, leave to erect four forges at Kimberworth, N.E. of Sheffield, two for reducing and two for fabricating iron, the ore being dug on the spot. In those days, and for centuries after, charcoal was used, until in 1619 Dud Dudley found a method of using coal (probably in the form of coke), but it was not until about 1735, that Abraham Darby completely solved the problem, by using coke. In 1796 there were 13 blast furnaces for smelting iron ore in the West Riding, which produced a total of 10,398 tons of pig iron ; in 1840 there were 32 furnaces (25 in blast), producing 56,000 tons ; in 1913, 21 furnaces (11 in blast), producing 302,695 tons, the total iron ore used being 871,587 tons. The Low Moor Co. (originally founded 1789) and the Bowling Iron Co., the latter of which ceased working in 1896, have obtained the highest reputation for their malleable iron.

The best steel in the world is produced in Sheffield. Before the year 1740 the steel produced there was chiefly of two kinds, blister steel and shear steel. The former was obtained by heating bars of imported (chiefly Swedish, charcoal-wrought) iron, sandwiched between layers of charcoal in a " cementation furnace," air being excluded, and the carbon combining with the iron to form steel. Shear steel is made by piling seven to ten plates of hammered blister steel one upon another, heating them to a welding heat and welding them under a hammer. In 1740 Benjamin Huntsman, born in Lincolnshire, who was at that time a clockmaker in Doncaster, finding blister steel unsatisfactory for clock springs, succeeded in inventing crucible cast steel, which he made at Attercliffe, Sheffield, by melting broken pieces of blister steel in clay crucibles, and pouring the molten metal into cast iron ingot moulds. It has been estimated that, in 1856, there were in Sheffield 105 converting furnaces for making cast steel, with 974 melting holes in them, and that in 1895 Sheffield used 14,000 clay crucibles weekly for steel melting. In 1856 the Bessemer process was invented by Sir Henry Bessemer, and enabled steel to be produced at a price so low that it was used for rails, girders, etc., instead of iron. Now the Bessemer process is being superseded by the Siemens-Martin and basic open-hearth processes, and progress is being made in melting steel by electricity.

The crucible cast steel trade continues to flourish, however, and among its latest developments is the making of high-speed tool steel, which, in tools, will continue to work at a red heat, and will take heavier cuts, at much greater speeds, in lathes and other machines, than ordinary steel. This tool steel contains varying quantities of

"Teeming" Crucible Cast Steel

one or more of the metals, tungsten, chromium, and vanadium. A rustless steel containing chromium is being made, and will be of great value in saving the labour of cleaning bright steel goods.

Sheffield "thwytels" (knives) were mentioned by Chaucer in 1386, and the city is best known for its pen-knives, pocket knives, table knives, and other cutlery, which includes bayonets. It also produces edge-tools, saws, files, hammers, spades and other gardening tools, steel wire and wire ropes, drills and plough plates. Railway material, such as locomotives, railway carriages and waggons, steel railway tyres rolled in one piece, axles, carriage and waggon springs and buffers, tramway rails and points and crossings are made in Sheffield. For marine purposes cast steel rudder-frames, propellers and shafts of all kinds, marine and other boilers, and the latest type of turbine drums are made there. The city is the greatest arsenal in England. Armour plates of such size that a single ingot may weigh 50 tons (though the finished weight is only about one third of the original), are made by three firms who can together produce 30,000 to 40,000 tons of them annually. Guns for the Navy up to 15-inch calibre and projectiles to correspond are also made, as well as forging-presses for forging and bending up to 12,000 tons power, rolling mills and engines, steam hammers, and shears for cutting steel bars and plates.

The number of men employed in Sheffield in 1911 in iron and steel manufacture was 20,158 ; in general engineering and machine-making, 13,439 ; in tool-making,

4485 men and 372 women ; in file-making, 3651 men and 1190 women ; in saw-making, 1243 men and 202 women ; in cutlery and scissor-making, 12,049 men and 2692

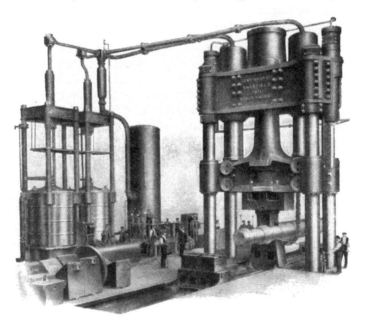

Six Thousand Tons High-speed Forging Press

women ; in wire-drawing, wire-making, and weaving, 1222 men and 51 women.

Sterling silver and electro-plate, together with Britannia metal (an alloy of tin, antimony, and copper) are important objects of Sheffield manufacture, employing about 10,000 persons before the war. The value of Sheffield electro-plate exported in 1911 was £1,086,400.

Woollen and worsted machinery is made chiefly in Leeds, Bradford, and Keighley, and about 20 other localities in the riding. Leeds is a great engineering centre, employing 21,205 men in general engineering and machine-making. Bradford, Halifax, Huddersfield, and in a less degree Rotherham and Wakefield, are also engineering centres, and Rotherham employs in addition 3170 men in iron and steel manufacture. The Great Northern Railway Co. has works at Doncaster for the manufacture of locomotives, waggons, and carriages, and the North Eastern Railway has carriage works and waggon repair works at York.

Pottery of an artistic character was formerly made in the riding, the most famous works being the Rockingham pottery at Swinton, near Rotherham, which about 1820 began to make porcelain of a highly decorative kind, and later produced a £5000 dessert service for William IV. It ceased working in 1842. More than 6000 persons are employed in the manufacture of glass bottles, chiefly in Barnsley, Leeds, Dewsbury, and Wakefield.

14. Mines and Minerals

The total value of the mineral output of the West Riding for 1913, excluding that of quarries under 20 feet deep, was £21,110,500. It is only surpassed by that of Durham (£22,267,865).

Alum-shale is worked in association with coal, chiefly at the horizon of the Stanley Main coal-seam of the Middle Coal Measures. The chief locality is near Lofthouse, north of Wakefield ; other localities are Thorpe Hesley near Rotherham, Smithies and Darton near Barnsley, and other pits in that district. The shale is used to make protochloride of iron and alum. In 1913 there was an output of 8741 tons.

Clay.—Fireclays are clays which resist high and long-continued heat without fusion, excessive shrinkage, or warping. Most of the fireclay worked in the West Riding is obtained from coal mines, and usually occurs as the underclay or " seatstone " of coal seams. It is found on the top of the Rough Rock of the Millstone Grit, beneath the lowest coal seam of the Lower Coal Measures, in pits about 5 miles W. of Sheffield, and is used for firebricks. The fireclay bed at this horizon is worked as far north as Huddersfield. At Grimescar, 2 miles N.W. of Huddersfield, the " Seggar Clay," above the Hard Bed Band Coal, is used for making the " seggars " or cases in which porcelain is fired. The best known in the riding is the Wortley fireclay, which occurs beneath the " Better Bed " coal, between Bradford and Leeds. The production of fireclay from coal mines of the West Riding in 1913 was 305,453 tons. In addition 1,029,571 tons of clay (partly fireclay but chiefly ordinary clay) were obtained from deep quarries, and used for the most part for bricks and tiles.

Coal is by far the most important mineral product of the West Riding. It was first mined in the riding

probably in Anglian, if not in Roman times, but most
likely on a very small scale, close to the outcrop of the
seams, and for local domestic use only. As long as
wood was plentiful, and no process of smelting iron by
means of coal had been discovered, there would be little
demand for coal. The digging of coal is recorded as
yielding some revenue to the estates of the Earl of
Lincoln near Pontefract, in 1240, and a coal miner
was killed by a fall of coal at Silkstone, in the reign of
Edward I. Even in the time of Henry VIII wood was
in most parts of England the ordinary fuel. The chief
coal seams and ironstone beds of the West Riding are
mentioned in the table on p. 80.

Until 1842, women and girls, and children as young
as from six to eight years of age, were employed underground in Yorkshire coal mines, but in that year an Act
of Parliament was passed, prohibiting all women and
girls from underground work, as well as boys below ten
years of age.

In 1855 the total output of the West Riding from
333 collieries was 7,747,470 tons, and in 1854, 21,030
persons were employed in the Yorkshire coalfield in coal-
mining, and 1164 in ironstone mining. In 1913 from
425 mines, 43,669,034 tons of coal (valued at the
pits' mouth at £20,570,955), besides fireclay, ironstone,
gannister, alum-shale, and iron pyrites were produced,
and 126,803 persons were employed underground, and
34,417 above ground.

Gannister, a rock consisting chiefly of quartz (silica),
with a little alumina, usually occurs as the seatstone

	Name of Coal-Seam, Etc.	Chief Area.	Quality.	Thickness.
Middle Coal Measure.	Shafton . . .	Thrybergh to Nostel.	Very fair.	3 ft. 6 in.
	Stanley Main . .	Wakefield to Featherstone.	Fair to good.	Av. 6 ft.
	Barnsley or Warren House . .	Sheffield to Darton and eastward.	Excellent.	4 ft to 9ft.
	Haigh Moor or Swallow Wood }.	{ N. of Wakefield and Pontefract. / S. of Barnsley.	Excellent.	4 ft.
	Tankersley Mussel-bed Ironstone	Tankersley, 4 miles S. of Barnsley.	Good in places.	3 ft.
	Strata 225 ft.			
	Thorncliffe Black Mine Ironstone	Chapeltown.		
	Parkgate (or Old Hards)	S. of High Hoyland.	Variable.	3 ft. to 4 ft.
	Middleton Main	E. of Batley Carr and Gomersall	Excellent.	3 ft. to 5 ft.
	Claywood Ironstone	Chapeltown.		
	Silkstone Coal	S. of Cawthorne.	Excellent.	4 ft. 6 in.
Lower Coal Measure.	Beeston Bed . .	Leeds.	Good.	6 ft.
	Black Bed Ironstone .	Low Moor, Bowling.		
	Gannister Coal . .	The Coalfield S. of Leeds.	Fairly good.	Av. 2 ft. 6 in.

The Chief Coal and Ironstone Beds of the West Riding

of coal seams, as already described. In 1913 West Riding coal mines furnished 95,717 tons, and deep quarries 19,201 tons of gannister, which is ground and made into fire-bricks, and linings of steel melting furnaces and converters.

Gravel and Sand.—Much of the gravel and sand in the riding comes from superficial glacial or post-glacial deposits, as for example at Bishopthorpe near York, Pollington, Heck, and Hensall Station, all near Snaith. At Glass Houghton, 1 mile S.E. of Castleford, the Permian quicksand is worked, and between Doncaster and Balby sand is obtained from Triassic sandstone. In 1913 from sandstone mines in West Riding, 8980 tons of sand, and from quarries in West Riding, 48,692 tons of gravel and sand were obtained, besides all that came from excavations under 20 feet deep.

Igneous Rock.—Only 800 tons of this product are recorded as quarried in the riding in 1913, from near Settle. Possibly the so-called Ingleton granite, which is really a grit, is meant. Igneous rock is very scarce in the riding. The chief mass is the diabase of Bluecaster, 4 miles N.E. of Sedbergh.

Iron Ore.—The chief ironstone beds of the West Riding are in the Coal Measures. They are, in descending order, the Swallow Wood, Lidgate, Tankersley or Mussel Bed, *Thorncliffe Black Mine, *Thorncliffe White Mine, *Claywood, *Black Bed Ironstone. The ore is in each case a clay ironstone, and occurs in thin beds or in nodules. The analyses of those marked above with an asterisk are very similar one to another, and show the

F

ore to be a carbonate of iron, with some carbonate of lime, magnesia, and manganese, and some argillaceous matter. The percentage of metallic iron is from 29 to 34. The ironstone worked commercially is the Black Bed Ironstone. It occurs in layers, the combined thickness of which never exceeds 22 inches. It is worked near Batley, Morley, and Low Moor near Bradford. In 1855 the West Riding produced 255,000 tons of ironstone; in 1868 the maximum of 785,000 tons; in 1913 only 28,269 tons from coal mines, and 590 tons from quarries. This falling off is due to the ores in the Lias, Oolite, and Cretaceous being more cheaply worked.

Iron Pyrites (iron disulphide) is picked out at some coal mines near Sheffield, and is used in making sulphuric acid. In 1913 25 tons were got in the riding.

Limestone.—In 1913 the output of limestone from the deeper quarries of the riding was 693,196 tons. The chief limestones quarried are those of Carboniferous and those of Permian age. The Carboniferous (including Yoredale) limestones are used for rough walling, but are chiefly quarried for lime-burning, cement-making, road metal, or iron smelting purposes. In the Dentdale district black marble has been quarried in Oliver Gill, and encrinital " marble " on Crag Fell and Snaizwold (Rise Hill). Black marble from Nidderdale was used in Fountains Abbey.

The Permian magnesian limestone affords excellent building stone. Much of York Minster is built of this limestone from Peter's Post, a part of Jackdaw Crag quarry, 2 miles S.W. of Tadcaster. Selby Abbey

is also built of magnesian limestone, and like Drax
Priory, its monks had quarrying rights near Tadcaster.
The keep of Conisbrough Castle, Tickhill church, Roche
Abbey (quarries close by), and part of Fountains Abbey,
are built of magnesian limestone.   Huddlestone quarry,
1 mile W. of Sherburn, supplied the material for
Huddlestone Hall and church, and together with North
Anston quarry near Shireoaks, stone for the Houses of
Parliament.   Other well-known quarries are at Brods-
worth, Cadeby, and Smawse (Bramham Moor).

Lead.—Lead mining appears to have been carried on
by the Romans in Yorkshire, and in 1903, at Lolley
Scarr mine, Pateley Bridge, 720 tons of dressed lead
ore were got.   The chief lead-mining districts were
Nidderdale, Wharfedale, and Airedale.   The ore (galena,
a lead sulphide), occurs in veins in Carboniferous rocks.
It is stated that 12,406 tons of lead ore were got in Yorkshire in 1856, but in 1913 none was mined in the West
Riding, and only 20 tons of ore in the North Riding.
The supply is not exhausted, but up to the outbreak of
war the price of lead had fallen so much, owing to the
importation of cheaper Spanish ore, that Yorkshire
mining became unremunerative.

Sandstone.—In 1913 the output of sandstone from
sandstone mines in the riding was 21,638 tons, and 404
men were employed.   In addition 742,533 tons of sandstone were got in the deeper quarries.   The West
Riding sandstones are chiefly of Carboniferous Age, and
some of them have a wide reputation as building stones.

The Millstone Grit is usually too hard and coarse to

furnish a good building stone, but Kirkstall Abbey is built of this stone from the old Bramley Fall quarries near Leeds. The older part of Fountains Abbey is built of Millstone Grit, so is the hydropathic establishment at Ben Rhydding near Ilkley. Millstone grit, called Bramley Fall stone, is quarried at Horsforth Woodside, N.W. of Leeds, and at Pool, E. of Otley, and is used for engine beds, dock works, and railways.

In the Lower Coal Measures the Soft Bed Flags are largely worked for flagstones near Gledholt, Huddersfield. The Elland Flagstone or Greenmoor Rock is in some places thinly bedded, furnishing hard and durable flags, in others it is thicker, and yields an excellent building stone, as at Well Hill quarry, near Wortley, in the Don Valley. It is largely worked on Elland Edge, at Rastrick near Halifax, at Swales Moor, North and South Owram, and between Thornton and Bradford. In the Middle Coal Measures the Birstal Rock affords very good building stone at Carlinghow and Howley Park near Birstal. The Thornhill Rock is the chief building stone of the Morley, Middleton, and Rothwell district. The Woolley Edge Rock is quarried at Hemingfield, 4 miles S.E. of Barnsley, and along the north side of Worsborough Dale, up to Barnsley, in both of which localities it is thick bedded and coarse. North of Wakefield it is finely grained, and is much sought after as a building stone. The Wickersley Rock and the Red Rock of Rotherham yield grindstones.

15. Shipping and Trade

The only seaport of the West Riding is Goole, though of course the produce of the riding is also shipped by other ports. Until 1820 Goole was a mere hamlet of no importance, but in that year the Aire and Calder Navigation Company obtained powers to make a canal between Goole and Knottingley, which was opened in 1826. The company spent large sums in making docks, and the port has continued to grow in importance. In 1905 the Lancashire and Yorkshire Railway Co. established steamship services between Goole and the Continent, and in 1912 had 25 steamers in use on these services. Goole, which has 3 miles of quays, ranks twelfth among the ports of the United Kingdom, with a value in exports and imports of £18,896,321 in 1911.

Exclusive of vessels engaged in coasting trade, the tonnage of British and foreign shipping entered at Goole in 1912 was 728,056, and cleared 720,773. Its chief exports are machinery, cotton yarn and manufactures, coal, stone, and woollen goods. The chief import is butter; but timber, seeds, fruit, and wool are also imported on a considerable scale. Shipbuilding is carried on here, and at Selby, where steel trawlers, etc., are built. According to the 1911 census, 1148 men were employed in ship and boatbuilding in the riding, inclusive of the county boroughs.

Goole

16. History

Although Julius Cæsar invaded Britain in 55 B.C., and again in 54 B.C., his expeditions did not result in the annexation of the country, and it was not until A.D. 43 that the Emperor Claudius commissioned Aulus Plautius to undertake the conquest of the island. By the year 47, the country south and east of a line joining the Severn and the Wash was subdued. Aulus Plautius was succeeded in 47 by P. Ostorius Scapula, who defeated the Brigantes in 49, and then made war on the Silures of South Wales, and the Ordovices of North Wales, under Caractacus, King of the Catuvellauni. Caractacus was completely defeated, and sought refuge among the Brigantes, the largest British tribe, who held almost the whole of what are now the six northern counties of England. He was by them surrendered to the Romans, who spared his life, took him as a captive to Rome, and ultimately released him. In the year 78 Julius Agricola, who had spent much of his life in Britain, was appointed governor, and according to the historian Tacitus, who was his son-in-law, subdued the Brigantes in 79. He was recalled to Rome in 85, and shortly afterwards the Brigantes rebelled. They again rose in the reign of the Emperor Hadrian, who came over to Britain in the year 120, and suppressed the revolt, ordering the construction of a wall of turf between Tyne and Solway, not only as a barrier against the Caledonians, but also as a means, by its garrison, of keeping the Brigantes under control.

In the year 208 the Emperor Severus came to Britain, and led his troops in two campaigns against the southern Picts and the Caledonians, returning worn out to Eboracum (York), where he died in 210, after causing the turf wall of Hadrian to be replaced by one of stone. The Emperor Constantius spent much time in Britain, and also died at Eboracum, and his son Constantine the Great was proclaimed emperor there—the first emperor who caused Christianity to be recognised by the State. From 360 to 367 the attacks of the Picts, Scots, and allied tribes, and of the Saxons, became more frequent and serious, and in the latter year they devastated a great part of Britain, though driven back and defeated by Theodosius in 368, when the Roman rule was re-established. Raids and attacks were, however, soon renewed, and in 410 the Emperor Honorius was in such difficulties that he bade the Britons defend themselves against the barbarians, which they did, at the same time establishing a government of their own. The Roman rule in Britain thus came to an end. Little appears to be known as to what happened in Yorkshire during the next century and a half, but by the end of that time the Angles had established two kingdoms north of the Humber—Deira, which extended from the Humber to the Tyne, and Bernicia, which lay between the Tyne and the Forth. Edwin, king of Northumbria, completed in 627 the conquest of the British kingdom of Elmet, a region of dense forest and bare moorlands, which according to some authorities answered roughly to the present West Riding of Yorkshire, while others hold that it comprised only about 72,000 acres between and including Leeds, Sherburn-in-Elmet, Barwick-in-Elmet, and Ledstone.

In 633 Edwin, who had been converted to Christianity by Paulinus, was defeated and slain in a battle at Heathfield (Hatfield, N.E. of Doncaster) against the pagan Penda of Mercia, allied with Cadwallon of Gwynedd (North Wales). Bernicia and Deira were separated, but union and separation recurred on many occasions later. Penda was defeated and killed by Oswy in 655 at Winwaed (believed to be Whin Moor, 5 miles N.E. of Leeds), and Oswy, by annexing Deira, became king of all Northumbria.

In 827 Northumbria acknowledged the supremacy of Egbert of Wessex, the first king of all England. In 867 the Danes took York, and in 876 their king, Healfdene, partitioned Deira among his followers. Danish kings ruled at York, though in 894 Alfred's supremacy was acknowledged. In 917 the men of York submitted to Æthelflaed, daughter of Alfred, and widow of Æthelred, ealdorman of Mercia, but in the following year Regnald, a Dane, came over from Ireland and seized the city. In 926 Æthelstan, king of England, drove out Guthfrith, king of York, gained in 937 a decisive victory over the Yorkshire and Irish Danes and the Scots at Brunanburh (a locality not as yet identified),[1] and maintained his rule over Yorkshire until his death in 940.

[1] There is a tradition that the battle was fought in the valley of the Axe, and it has also been suggested that Burnley may coincide with Brunanburh.

Alternately Danish and English kings held sway, and Deira and Bernicia were now united in the earldom of Northumbria, now separate.

In 1066 Tostig, Earl of Northumbria, in alliance with Harold Hardrada, king of Norway, took York, but was defeated and slain by Harold, king of the English, on 25th September at Stamford Bridge, on the Yorkshire Derwent. Nineteen days later Harold was himself defeated and killed at the Battle of Senlac (Hastings), by the army of William the Conqueror.

Several years passed before the conquest of England was completed. Even by the summer of 1068 no step had been taken practically to subjugate Yorkshire. The earls, Edwin and Morkere, who had submitted to William, revolted and were supported by the people of Yorkshire, but when they found that William had seized and fortified Warwick, they submitted a second time, and York opened its gates on the approach of the Conqueror. He at once built a castle between the Ouse and Foss to overawe the city, but in January 1069, the citizens of York again revolted, killed Robert Fitz-Richard and many of his companions, and laid siege to the castle. William came, slaughtered or dispersed the besiegers, and built a second castle on the opposite bank of the Ouse. In September 1069 the English, assisted by a Danish fleet, attacked and took the castles of York, sparing but few of their defenders. They then dispersed. William returned to the city, much of which had been burnt, and ordered the castles to be repaired. He then exacted a terrible vengeance by relentlessly harrying Yorkshire and the neighbouring shires. All who withstood him were slain, houses and their contents and property of every kind were burnt and destroyed. In the great Domesday survey of 1086, a large part of the West Riding (exclusive of Craven) is returned as "waste," namely the tract extending from Armley to Gargrave, and from Holmfirth to Adel. Part of Craven was also waste.[1]

The Scots after their victory at Bannockburn in 1314 made several raids on Yorkshire between 1316 and 1321, sacking houses and churches, including Fountains Abbey and Bolton Priory. In 1319 the Archbishop of York, at the head of a motley array of 10,000 men, attacked a Scotch army of 15,000 on the tongue of land between Swale and Ure, opposite Myton, but was defeated with enormous loss. Edward II, by the favour he showed to worthless favourites, drove his cousin, Thomas, Earl of Lancaster, and others, into rebellion. The earl, finding that the king's forces were superior to his own, endeavoured to lead his troops towards Dunstanburgh in Northumberland, but was met at Boroughbridge by a force under Sir Andrew de Harcla, acting for the king. A battle took place, 16th March 1322, on and near the bridge, in which the Earl of Hereford was killed. Next day the Earl of Lancaster was captured, and on 22nd March was executed on a hill outside Pontefract. In 1400 Pontefract Castle became the

---

[1] The word "waste" in Domesday Book usually meant that there were practically no oxen, and little or no live stock; but does not necessarily mean that there were no inhabitants.

prison of Richard II, who met his death there.   It was
the general belief that he was murdered, and this view,
and the unpopularity of Henry IV led to a rebellion
in 1405, headed by the Earl of Northumberland and
Thomas Mowbray, Earl Marshal of England, who were
joined by Richard Scrope, Archbishop of York.   The
Archbishop, at the head of twenty thousand men, en-
camped on Shipton Moor, 5½ miles N.W. of York.   Here
the Earl of Westmorland treacherously captured Mowbray
and the Archbishop, and delivered them to the king, who
had them both executed near York without proper trial.

In 1408 the Earl of Northumberland and Lord
Bardolph made another attempt to dethrone Henry
IV.   At the head of a Scotch force, and joined by others,
they were defeated by Sir T. Rokeby at Bramham Moor,
3½ miles W.S.W. of Tadcaster.   The Earl was killed
and Lord Bardolph taken prisoner, but he died on
the field.   In the Wars of the Roses Richard, Duke of
York, with not more than 5000 men, unwisely left the
shelter of Sandal Castle, and gave battle to a Lancastrian
army of 20,000 men on Wakefield Green, on the 30th of
December 1460.   He was defeated and slain with over
2000 of his troops.   The triumph of the Lancastrians
was short-lived.   On March 29th, 1461, Edward, the
new Duke of York, with 40,660 men, confronted a
Lancastrian army of 66,000 men.   The Yorkists were
at Saxton, and the Lancastrians at Towton, 2 miles to
north of them, the little River Cock was to the west.   A
driving snowstorm blew in the faces of the Lancastrians,
and caused their arrows to fall short.   The battle, the

bloodiest fought on English ground, resulted in their overwhelming defeat. The number of killed of both parties was between 28,000 and 38,000 men, many Lancastrians being drowned in the stream, over which there was only one inadequate bridge.

In 1536 the suppression of the monasteries and other unpopular measures of Henry VIII caused a rebellion in Yorkshire, known as the " Pilgrimage of Grace." Its leader was a lawyer named Robert Aske, who, with 35,000 men, confronted the king's forces of 8000 under the Duke of Norfolk, on the opposite side of the river Don at Doncaster. A conference was held, the king's pardon and a parliament at York were promised, and the rebels dispersed ; but when another rebellion occurred the following year, the king made it an excuse for executing Aske, though the latter had done his best to discourage the rising.

In 1537 the Council of the North, with its head-quarters at York, was appointed by Henry for the control of the Scottish border, and of England north of the Humber. Its arbitrary jurisdiction was abolished in 1641. In 1569 a rising took place in Yorkshire and Durham, with the object of releasing Mary Queen of Scots from Tutbury Castle and restoring the Roman Catholic religion, but it was put down, and hundreds of those who had participated in it were hanged.

In the Civil War the Royalist Earl of Newcastle, with 6000 foot and 200 horse, crossed the Tees on December 1st, 1642, and advanced into Yorkshire. The Parliamentarian Lord Fairfax, and his son, Sir Thomas, held the manufacturing towns of the West Riding, which were on the popular side, together with Selby and Cawood on the Ouse, and Hull, which was under the treacherous Sir John Hotham. Lord Fairfax attempted to hold the line of the Wharfe at Tadcaster with 800 men, but after five hours' fighting on December 7th, withdrew during the night to Selby. Towards the end of December Sir Thomas Fairfax went to the assistance of Bradford, and from there advanced against Leeds, where he defeated Sir W. Savile, and took the town on January 23rd, 1643. Towards the end of January he himself returned to Selby.

Owing to the treachery of the Parliamentary commanders at Scarborough and Hull, Lord Fairfax left Selby on March 29th and marched to Leeds. Sir Thomas, who protected his flank, suffered a defeat on Seacroft (Whin) Moor, the supposed site of the Battle of Winwaed, but on May 21st, with only about 1200 men, he was able to take Wakefield, although it was defended by 3000 foot and 7 troops of horse. At last the Earl of Newcastle marched against the Fairfaxes at Bradford with a force of 10,000 men. With barely 4000 men they met him on Adwalton Moor, 4 miles S.E. of Bradford, and were defeated on June 30th. Bradford was taken by Newcastle on July 2nd, but the Fairfaxes escaped through Leeds and Selby to Hull.

Newcastle began the siege of Hull on September 2nd, but it was so stoutly defended that, after little more than a month, he relinquished his attempt. On April 11th, 1644, the Fairfaxes stormed Selby, which was garrisoned by 3300 men under Lord Bellasis, and 81 Royalist officers and 1600 men were made prisoners. In September 1643 the English parliament signed a treaty with the parliament of Scotland, by which a Scottish army was to assist the English. This army, under the Earl of Leven, was in county Durham, facing that of the Earl of Newcastle, when news of the fall of Selby arrived. Newcastle advanced by forced marches to save York for the king. Leven advanced with even greater rapidity, and joined the army of Lord Fairfax at Tadcaster, on April 19th, 1644. They began the siege of York, and were joined on June 2nd by the army of the Earl of Manchester, their united forces being 22,000 foot and 4100 horse. On June 30th the besiegers heard that Prince Rupert, with 20,000 men, was approaching to relieve the city, and they accordingly retired to Marston Moor, 7 miles W. of the city. Rupert entered York on July 1st, took over from Newcastle the supreme command, and with his usual impetuosity decided to give battle. His forces amounted to about 22,000 men, those of the allies to 15,000 foot and 9000 horse, besides gunners. Oliver Cromwell commanded Manchester's Horse. On the evening of July 2nd, the great battle of Marston Moor took place, and resulted in a decisive victory for the Parliamentary forces. Three thousand, or according to some accounts 4000 Royalists were slain, as compared with 1500 of the Parliamentarians, and 1500 Royalists were taken prisoners. York surrendered to the Parliamentary forces on July 16th ; Tickhill Castle, July 26th ; Sheffield Castle, August 11th ; Knaresborough Castle, December 20th, 1644. Next autumn Sandal Castle fell on October 2nd, and Skipton Castle, after a three years' siege, on December 21st, 1645. Pontefract Castle, twice besieged in 1644–5, taken in the latter year, and lost again in 1648, was again besieged in 1648–9, and finally surrendered on March 22nd, 1649. All these castles, with the addition of Cawood Castle (taken in May 1644), were ordered to be made untenable. With the final capture of Pontefract the Civil War in Yorkshire came to an end.

17. Antiquities—(a) Prehistoric

Palaeolithic implements do not occur in the riding, but Neolithic flint and stone implements have been found abundantly in many places, especially on the moors around Halifax, Huddersfield, and Keighley. Rombalds (Rumbles) Moor, near Ilkley, however, has yielded the greatest number.

The " Devil's Arrows," three great monoliths of Millstone Grit of Neolithic times, are about a quarter of a mile west of Boroughbridge. They are from 18 to 22½ feet in height, and stand in a north and south line. Rombalds Moor is a heather-clad plateau, 1000 to 1100 feet above sea-level, between the valleys of the Wharfe and Aire, where barrows, stone circles, encampments, etc., are numerous. A mile and a half S.E. of Ilkley is an earthwork, more than 600 feet long from E. to W., its eastern portion forming an irregular parallelogram, surrounded by a bank of earth and stone

8 feet in diameter and 2½ feet high. One end of this
great earthwork terminates in a circle exactly like some

Devil's Arrows, Boroughbridge

of the barrows of the Bronze Age, and it is probably of
that period. About half a mile S.E. of this is a

G

cairn, called the Little Skirtful of Stones, about 60 yards in circumference, and 7 feet high, and half a mile S.S.E. of it is a larger cairn, the Great Skirtful of Stones, 85 feet in diameter and 5 to 6 feet high, composed of boulder stones. These sepulchral barrows are possibly of the Neolithic Age. 130 yards S.S.E. of Grubstones Shooting House is another barrow of a different type, consisting of a stone circle 31 ft. 6 in. in diameter, near the centre of which some bones and ashes and a flint spear-head were found. Eighty yards E. of the Great Skirtful is a circular earthwork, 93 feet in diameter ; and ¼ mile S.E. of the Skirtful another, 80 feet in diameter. These have been called camps, as no traces of burial have been discovered in them. A mile S.W. of Great Skirtful is Horncliff Shooting House, 1 furlong N. of which is a stone circle, 43 feet in diameter, the stones varying from 3 to 5 feet in height, with a small circle of 7 stones in its centre. Another (12-stone) circle called the Twelve Apostles occurs ¼ mile S. of Lanshaw Delves. On Baildon Moor, ½ mile N.N.W. of Hope Hill Farm, is a barrow consisting of an earthwork circle about 50 feet in diameter, in the centre of which a cinerary urn was found. Northward of this, on Penythorn Hill, an urn containing calcined bones, a flint flake, a small bronze weapon like a knife-dagger, and a perforated piece of bone, was unearthed in 1904. This shows that these burials are of the Bronze Age, and the stone circles are probably contemporaneous.

"Cup and ring markings" are a rude kind of sculpture, consisting of small circular cup-like cavities, usually surrounded by incised concentric circles. Neighbouring circles are sometimes united by two parallel grooves, with transverse lines forming a ladder-like pattern. From the association of such markings with cinerary urns of Bronze Age on Ayton Moor, near Scarborough, it is clear that they are of the same period. The best and most easily accessible example was *in situ* about 100 yards west of the Panorama Rock, 1 mile S.W. of Ilkley, but it has been removed to a place opposite St Margaret's Church in Ilkley. Several other examples occur, *e.g.* 1¼ miles S. of Ilkley, between the roads to Keighley and Bingley.

On the flat summit of Ingleborough is a camp, probably British, 15 acres in area, surrounded by a low rampart of stone, and containing the horseshoe-shaped foundations of nineteen hut dwellings. Some earthworks near Sheffield are of interest. One of them is an oval camp, 133 yards long by 103 wide, lying ¾ mile W.S.W. of Wincobank Station. A similar but smaller oval, popularly called Cæsar's Camp, lies 3 miles N.N.E. of Wincobank Camp, in Scholes Coppice, Wentworth Park. Neither of these camps is likely to be Roman. An embankment called the Roman Rig (ridge), with a ditch on its southern side, can be traced at intervals N.E. from Wincobank Camp, through Greasbrough towards Swinton, and is probably British. Another embankment, also called Roman Rig, runs N.N.E. through Wentworth Park to Mexborough, and has been considered by some authorities to mark a Roman road.

18. Antiquities—(b) Roman

The Romans were fully aware that to keep in subjection a country conquered by the sword, both fortresses and good roads are essential. Their roads are usually of undeviating straightness, and are often raised above the neighbouring land. In many cases they have been incorporated in modern roads. Ermine Street led north from London to Lincoln, and entered the West Riding at Bawtry, whence it ran through Danum (Doncaster, N.W. of which, between Barnsdale Bars and Bodles, it is well preserved) to Legiolium (Castleford), thence to Calcaria (Tadcaster), from which a road led to Eboracum (York). A little W. of Tadcaster Ermine Street continued northward to Isurium (Aldborough), being joined at Providence Green, $1\frac{1}{2}$ miles N. of the Nidd, by another road from York. At Aldborough it left the riding, proceeding to Catterick, Bishop Auckland, and beyond.

Eboracum became an imperial residence, the seat of justice, and the headquarters first of the 9th, and for nearly 300 years, of the 6th Legion. It was one of the chief, if not the chief Roman city in Britain. The Roman fortress was on the left bank of the Ouse. It formed a rectangle, measuring 540 yards from N.E. to S.W. by 470 yards from N.W. to S.E. enclosing over $52\frac{1}{4}$ acres, and surrounded by a wall, 4 to 5 feet thick. At each of the four angles there was probably a tower, besides many minor towers in the walls. The great tower at the N.W. angle still exists in the grounds of the Yorkshire Philosophical Society as the "Multangular Tower," crowned by a medieval superstructure. It had ten sides. The fortress had four principal gates, one on each side, and the remains of one of them has been discovered beneath Bootham Bar. Numerous remains of Roman inscriptions, altars, sarcophagi, sculpture, pottery, glass, fibulae, bracelets, finger-rings, pins, coins, and other objects have been found, and most of them are to be seen in the museum of the Yorkshire Philosophical Society.

Iseur, the capital of the Brigantes, became, under the name of Isurium (Aldborough), an important Roman station. Its walls, 11 to 16 feet thick, were 6440 feet in circumference, forming an irregular rectangle of about 2000 feet by 1275 feet, and enclosing 60 acres. Its tesselated pavements are noteworthy, and many of its Roman remains are preserved in the local museum.

Riknild Street from Worcester is believed to have crossed the Don near Aldwarke, and running approximately parallel to Ermine Street by Normanton and Woodlesford, and the Roman camp at Minskip, joined Ermine Street N.W. of Aldborough. From Aldborough a road ran to Olicana (Ilkley), where was a camp measuring about 300 feet by 480 feet, encompassing the present parish church. From Tadcaster a road passed the Roman station of Burgodunum, north of Adel, also leading to Ilkley. Ilkley and Ribchester (Bremetennacum) were connected by a road passing near Skipton, which was commanded by the fort of Calagum

(Burwen Castle, Elslack), consisting of a smaller and earlier fort about 380 feet square, enclosed within another 603 feet by 406 feet, with a stone wall 8 feet 6 inches thick. Another road from Ilkley crossed Rumbles Moor, and led via Denholme, Sowerby, Blackstone Edge, and

Roman Road, Blackstone Edge

Littleborough to Mancunium (Manchester). On Blackstone Edge it is better preserved than any other Roman road in the riding, and is about 18 feet wide, paved with rectangular stones, and with a remarkable central stone channel or trough, 14 inches wide by 5 to 8 inches deep.

Another Roman road from Manchester passed by

Failsworth and Delph, a mile and a half N.N.E. of which it reached the Roman camp at Castle Shaw, the fort of which measured about 400 feet square. Thence it ran north-eastward to Cambodunum (Slack, 4 miles W. of Huddersfield), where there was a walled camp, a building with a hypocaust (hot air chambers and flues) and other remains. It is supposed that the road from Slack led via Leeds to York.

A road called Batham Gate, from Aquae (Buxton) led north-eastward by the camp at Anavio (Brough), and over Stanage Edge, by the Long Causeway to (or near) the Roman fort at Templeborough between Sheffield and Rotherham.[1] It probably joined Riknild Street. From near Askrigg another Roman road crossed Wether Fell and Cam Pastures to Over Burrow near Kirkby Lonsdale.

19. Antiquities—(c) Anglo-Saxon

Anglo-Saxon antiquities of a portable character do not appear to have been recorded as occurring in the riding west of Ripon, Leeds, and Pontefract. But near the church at Goldsborough a hoard of Anglo-Saxon and Kufic (Asiatic) silver coins, with silver brooches and bracelets, was found in 1859, and is now in the British Museum, as is a gold finger-ring of Æthelswith, sister of Alfred the Great, found between Aberford and Sherburn. A hoard of 10,000 Saxon coins was found at St Leonard's Place, York, in 1842.

---

[1] Recently excavated and destroyed.

Pre-Conquest Crosses, Ilkley

Anglian and Anglo-Danish sculpture has been recorded from at least 42 (possibly 66) sites in the West Riding, extending as far west as Burnsall, Gargrave, and Rastrick. It is chiefly in the form of stone crosses, usually adorned with interlaced or straight-lined patterns, leaf, and fruit scrolls, tree scrolls, bird scrolls, figures of beasts, or snakes, and finally human figures.

In the churchyard of the parish church, Ilkley, are three celebrated shafts, probably of the eighth or ninth century. The tallest has on one side four panels showing the four evangelists ; on the opposite side Christ enthroned, and three panels of grotesque beasts ; the other two sides, have spiral scrolls. A cross head of similar date has been affixed to this shaft. Other remains are in the Ilkley Museum.

Numerous Anglo-Saxon fragments are preserved at Dewsbury parish church. At Leeds, in the parish church, is a great cross, probably early tenth century, the shaft (partly restored), with scrolls, knots, figures of saints, and a figure supposed to represent the legendary Wayland the Smith. Other fragments are in the museum of the Philosophical Society.

At Otley church there are portions of the shaft and head of a beautiful Anglian cross, also of another shaft with sculptured dragons and other fragments.

Ten fragments, probably of the ninth century, some of them grave-slabs with inscriptions in Anglian runes, are to be seen at Thornhill church, near Dewsbury. The churches of Birstall, Burnsall, Collingham, and Kirkby Wharfe, also contain fragments of pre-Conquest sculpture.

20. Architecture—(*a*) Ecclesiastical

A preliminary word on the various styles of English architecture is necessary before we consider the churches and other important buildings of our Riding.

Pre-Norman or, as it is usually, though with no great certainty termed, "Saxon" building in England was the work of early craftsmen with an imperfect knowledge of stone construction, who commonly used rough rubble walls, no buttresses, small semicircular or triangular arches, and square towers with what is termed " long-and-short work " at the quoins or corners. It survives almost solely in portions of small churches.

The Norman Conquest started a widespread building of massive churches and castles in the continental style called Romanesque, which in England is usually called " Norman." They had walls of great thickness, semicircular vaults, round-headed doors and windows, and massive square towers.

From 1150 to 1200 the building became lighter, the arches pointed, and there was perfected the science of vaulting, by which the weight is brought upon piers and buttresses. This method of building, the " Gothic," originated from the endeavour to cover the widest and loftiest areas with the greatest economy of stone. The first English Gothic, called " Early English," from about 1180 to 1250, is characterised by slender piers (commonly of marble), lofty pointed vaults, and long, narrow, lancet-headed windows. After 1250 the windows became broader, divided up, and ornamented by patterns of tracery, while in the vault the ribs were multiplied. The greatest elegance of English Gothic was reached from 1260 to 1290, at which date English sculpture was at its highest, and art in painting, coloured glass making, and general craftsmanship at its zenith. This is known as the " Geometrical" or " Early Decorated " period.

About 1300 the structure of stone buildings began to be overlaid with ornament, the window tracery showed rich double curvature and ogee arches, the vault ribs were of intricate patterns, the pinnacles and spires loaded with crocket and ornament. This latter style is known as " Curvilinear " or " Late Decorated," and came to an end with the Black Death, which stopped all building for a time.

With the changed conditions of life the type of building changed. With curious uniformity and rapidity the style called " Perpendicular "—which is unknown abroad—developed after 1360 in all parts of England and lasted with scarcely any change up to 1550. As its name implies, it is characterised by the perpendicular arrangement of the tracery and panels on walls and in windows, and it is also distinguished by the flattened arches and the square arrangement of the mouldings over them, by the elaborate vault-traceries (especially fan-vaulting), and by the use of flat roofs and towers without spires.

The medieval styles in England ended with the dissolution of the monasteries (1530–1540), for the

Reformation checked the building of churches. There succeeded the building of manor houses, in which the style called " Tudor " arose—distinguished by square-headed windows, mullions and transoms, level ceilings, and panelled rooms. The ornaments of classic style

Bardsey Church

were introduced under the influences of Renaissance sculpture and distinguish the " Jacobean " style, so called after James I. About this time the professional architect arose. Hitherto, building had been entirely in the hands of the builder and the craftsman.

Of pre-Norman work in our riding we have the tower, nave (now S. aisle), and chancel of Kirk Hammerton church, near York ; the tower of Bardsey church, and that of Monk Fryston, near Pontefract. The nave, chancel arch, and lower part of the tower of Ledsham church, also near Pontefract, are Anglo-Saxon ; and in Laughton-en-le-Morthen church, near Rotherham, a doorway in the N. wall and the walls at the N.W. end belong to the same period.

The West Riding has some very good examples of Norman architecture. The very interesting church of Birkin, 7 miles S.W. of Selby, is Norman (except the Decorated S. aisle), and terminates eastward in a semi-circular apse. Adel, 5 miles N.N.W. of Leeds, is a small but fine Norman church, consisting of nave and chancel only, with a beautiful S. porch recessed in five orders, and a good chancel arch. In Bardsey church, already mentioned, the N. arcading of the nave and the S. doorway are Norman. In Brayton, near Selby, the chancel arch, lower part of the tower, and S. doorway are of this period. Leathley has an early Norman tower, Selby Abbey church, to be described later, shows some important Norman features, and the crypt of York Minster is chiefly Norman. Both St Denis and St Margaret's, Walmgate, York, have beautiful Norman doorways.

Of Transitional Norman work there is also abundance. In Thorpe Salvin church, 11 miles E.S.E. of Sheffield, the arches of the tower and chancel, the S. doorway, and the very remarkable font are Transitional Norman, and in Laughton-en-le-Morthen the N. aisle, chancel, and piers of the N. nave arcade, as is much of Throap-

South Porch, Adel Church

ham (St John's), a mile distant.   Conisbrough church
has a Transitional Norman nave arcade, chancel arch,
and S. doorway, and Campsall an interesting W.
tower, chancel arch, and transept arches of this period.
Hatfield, Sherburn-in-Elmet, Bardsey, and Healaugh
churches all show noteworthy work of this style.

The finest examples of Early English at York Minster,
Ripon Minster, Selby Abbey, and Nun Monkton Priory
church are described under Religious Houses.   The
church of Adlingfleet, near the junction of Ouse and
Trent, is chiefly Early English, as is the tower and west
bay of the nave of Snaith church.   To the same period
belong the chancels of Drax and Sherburn-in-Elmet,
the nave, transepts, and S. aisle of Church Fenton,
and the arch of S. aisle of chancel and Slingsby chapel,
of Knaresborough.

Apart from the Decorated work in York and Ripon
Minsters and Selby Abbey, to be described later, we have,
as showing good examples of this period, Walton near
Wetherby (except the tower), and Methley near Wakefield (except the tower and Waterton chantry).   In
Royston church near Barnsley, the nave, clerestory,
and chancel ; in Darfield, the nave and aisles ; and in
South Anston, almost all except the tower and Transitional Norman arcade, belong to this period.   Here
may be mentioned the chantry chapel of St Mary, on
the bridge at Wakefield, probably built shortly after
1342, but rebuilt in 1847.   The original west front is at
Kettlethorpe Hall, 2 miles S. of the bridge.

Churches in the Perpendicular style are numerous in

the Riding. Parts of Wakefield and Sheffield cathedrals, both originally parish churches, are in this style. In Sheffield Cathedral are the tombs of the fourth Earl of Shrewsbury with his effigy, and those of his two wives, and of the sixth Earl, the custodian of Mary Queen of Scots. Other Perpendicular churches are Halifax parish church ; Harewood, with the effigies of Sir W. Gascoigne and his wife (he was the judge who committed the prince, afterwards Henry V, to prison) ; Rotherham and Ecclesfield, each with transepts and a central tower ; Bolton-by-Bowland, with the tomb of Sir Ralph Pudsay, who sheltered Henry VI after the battle of Hexham ; Skipton (partly Decorated), with the monuments of the Cliffords ; Tickhill, with the effigies of William Fitzwilliam (d. 1478) and his wife ; Almondbury, with a remarkable inscription of 1552 round the nave ; Bolton Percy, a fine church, with a monument of Ferdinando, Lord Fairfax, who commanded part of the Parliamentary army at Marston Moor. In Methley, the Waterton chantry is Perpendicular, and contains fine effigies of Sir R. Waterton (d. 1424) and his wife ; and Lord Welles (killed at the Battle of Towton in 1461), and his wife. The neglected chantry chapel on the bridge at Rotherham is Perpendicular, as are Penistone and Darton. Silkstone, though chiefly of this style, has Norman piers in the chancel, and remarkable buttresses pierced by gargoyles.

Before describing the religious houses of the West Riding, we may devote a few words to the monastic orders to which they belonged.

The Benedictines, or Black Monks, lived according to the rule of St Benedict, who was born at Nursia (Norcia) in Italy, about A.D. 480. They were not strictly an Order with centralised government, but consisted of independent communities. Their rule was not of great austerity. They usually devoted four or five hours of the day (more on Sundays) to religious duties, four hours to reading, about six hours to work, chiefly field work, and seven or eight hours to sleep.

The Cluniac Order was an offshoot of the Benedictine, and took its name from the abbey of Cluny, near Mâcon in France, founded in 910. Under its rule manual labour gave place to religious observances, and the abbot of the original monastery had great power over all other houses of the Order.

The name of the Cistercians (Grey or White Monks) is derived from that of the abbey of Cîteaux (*Cistercium*) founded near Dijon in 1098. The Cistercian was the chief offshoot from the Benedictine Order. They endeavoured to go back to the original characteristics of St Benedict's rule, and made a point of field work, in which they were assisted by lay brothers, who were uneducated labourers.

Monks were not necessarily priests, but regular canons were essentially priests. The Augustinian Canons (Austin Canons) followed the so-called rule of St Augustine, drawn up towards the end of the eleventh century. They resembled monks in living in communities, but served as priests of the churches under their patronage.

H

Selby Abbey

Unlike monks and canons regular, friars were not necessarily attached to a religious house in one definite locality, but belonged to a wider province or order.

Selby Abbey, Triforium and Clerestory

There were four great orders of mendicant friars, Franciscans, Dominicans, Carmelites, and Augustinians.

There were five houses of Benedictine monks in the riding, the most important being Selby Abbey, St Mary's Abbey at York, and Monk Bretton Priory.

The abbey church of Selby (built of magnesian limestone) has a W. front which is transition Norman below, and Early English above, with low, flanking towers, a Norman N. portal, and transepts, one Norman, but the southern one (together with the central tower), built since the fire in 1906. The interior of the nave passes from Norman on the E. to Early English on the W., and from Norman in the main arcade to Early English in the triforium and clerestory. The choir is Decorated.

William Rufus laid the first stone of the original St Mary's Abbey, York, which terminated eastward in seven apses, of which the foundations only remain. It was destroyed by fire in 1137, and the church of which the picturesque ruins still exist was begun in 1271. It was cruciform, the nave of eight and choir of nine bays, of almost equal length; the transept had an eastern aisle, and there was a central tower. The chief existing remains are the wall of the N. aisle, with eight Decorated Geometrical windows. Between each pair of windows are two blind lancets, and the stumps of the ribs of the groined roof. Beneath the windows is a Geometrical blind arcade. Part of the W. end of the nave, the arch at the E. end of the N. aisle, and part of the N. transept still remain, as well as the abbey gatehouse and hospitium, and the foundations of most of the original buildings.

Of Monk Bretton Priory, originally Cluniac, $1\frac{1}{2}$ miles E. of Barnsley, little remains but the Perpendicular gatehouse (now a barn), and some walls with Decorated windows.

There were two houses of Benedictine nuns in the

riding, the chief of these being the priory of Nun Monkton, at the confluence of the Ouse and Nidd. The beautiful priory church has a Transitional Norman W. doorway and three others on the S. side. The interior is one of the finest examples of the Early English

St Mary's Abbey, York, North Aisle

style in the riding. There are no aisles, and the chancel is modern.

The Cistercian monks possessed four abbeys in the riding, viz., Fountains, Kirkstall, Roche, and Sawley.

Fountains Abbey is one of the most beautiful and most beautifully situated abbeys in the kingdom, and it is also notable in that not only the church, but many of the other monastic buildings are preserved. It lies three miles S.W. of Ripon, on the Skell, some of the buildings being actually supported on arches or tunnels over the stream.   The nave of the abbey church is Norman, of eleven bays, divided from the aisles by ten massive circular piers on each side, with slightly

Fountains Abbey

pointed arches ; the windows, both of aisles and clerestory, are round-headed Norman.   At the crossing of nave and transepts was a central tower which has fallen, but between 1494 and 1526 a tower, nearly 170 feet high, was built at the north end of the N. transept. East of the transepts is the presbytery, in Early English style, with lancet windows.   The most beautiful feature of the church is the "Chapel of the Nine Altars." Early English in style, with lofty lancet windows, its vaulted roof was supported near the centre by two beautiful slender octagonal pillars about 50 feet high, each originally surrounded by eight marble shafts.

Fountains Abbey, Cellarium

The cellarium — incorrectly called cloisters — was over 40 feet in width, with a vaulted roof supported by a central row of polygonal pillars.

Kirkstall Abbey is 3 miles N.W. of Leeds. The church is Transitional Norman, with the exception of the east end of the choir and the upper part of the central tower, which are Perpendicular. The transepts north and south have each three chapels on the eastern side ; the choir is very short.   The monastic buildings are more severe in style than those at Fountains.

Roche Abbey is charmingly situated 7½ miles E.S.E. of Rotherham.   It is built of magnesian lime-stone from the celebrated quarries close by.   The chief remains are the eastern sides of the transept, with a portion of their chapels and of the choir—all Transitional Norman, with pointed arches, but round-headed windows. The gatehouse is Decorated.

The ruins of Sawley (Salley) Abbey, 3½ miles N.E. of Clitheroe, are of little architectural interest.   It was largely built of shale and boulders, but partly of sandstone.   The church was 185 feet long, of which the nave occupied only 40 feet.   The nave is Norman, the choir Perpendicular.   The arrangement of the various monastic buildings can still be made out.

The Cistercian nuns had five priories in the riding. Of these the only remains are at Kirklees, 2 miles E.S.E. of Brighouse ; and a beautiful Norman door-way of Syningthwaite Priory, in a farmhouse, 3½ miles E. of Wetherby.

The Cluniac monks had a priory at Pontefract (of which no remains exist), as well as that at Monk Bretton. Of the priory of the Cluniac nuns at Arthington, there is nothing now standing.

At Drax, Healaugh Park, Nostell, and Bolton Priory (generally called Bolton Abbey), were priories of the Austin Canons.   Amid beautiful river scenery Bolton Priory stands in ruins, with the exception of the nave of the church, which has been restored.   It shows an interesting mixture of styles. The south-west part of the nave is Transitional Norman, the rest, including the N. aisle, Early English. In front of the west end is a Perpendicular tower, never completed. The transepts are partly Decorated, partly Transitional Norman, with some Early English on the W. side of the N. transept. The east end and upper part of the choir is Decorated, the lower part has a Transitional Norman blind arcade of interlacing semi-circular arches.

The Gilbertine monks had St Andrew's Priory at York. The Knights Templars had six, and the Knights Hospitallers three preceptories in the West Riding, including York. There were five friaries in York, two at Doncaster, and one each at Pontefract, Knaresborough, and Tickhill. No less than 40 medieval hospitals existed, which helped the poor and infirm, as well as pilgrims, lunatics, and lepers. That of St Leonard at York was the largest and wealthiest. Its ruins are near St Mary's Abbey.

The splendid cathedral church of York has an interior length of 486 feet, and its nave, 99 feet high, consists of eight bays, the choir of nine. The central and two western towers are of almost equal height (198 and 196 feet). The building is mainly of three periods and styles of architecture. The main transepts which have aisles on both sides and the largest triforium in England, are Early English. That to the south contains the most used portal of the church, its façade much loaded with ornament ; that to the north has a group of lancet windows, 50 feet in height, celebrated as the " Five

Sisters.'' The nave, the lower part of the west front, the beautiful octagonal chapter-house, and the vestibule connecting .it with the N. transept are Decorated (Geometrical). The West window and the tower windows on either side of it are Decorated (Curvilinear). The choir, with its great East window, 78 by 32 feet, the central tower, and the upper parts of the western towers are Perpendicular. In the crypt are circular Norman columns and piers, one of them with the lattice-work pattern on its shaft, which occurs in Durham Cathedral and Selby Abbey church. There are also remains of a wall with herring-bone work, supposed to be Saxon. The greatest glory of the minster, however, is its magnificent stained glass.

Ripon Cathedral has a Saxon crypt and on the S. side of the choir a Norman apsidal building. It was probably a chapel of a Norman cathedral, but is now used as a chapter-house and vestry, and has a Norman crypt beneath it. The nave, as built by Archbishop Roger, between 1154 and 1181, was Transitional Norman, and had no aisles. The lower part was a blank wall ; above it the triforium was a mere passage in the thickness of the wall, with a lofty pointed arcade in front of it ; the clerestory alone had windows, small lancets and large round-headed ones. Of this nave two portions on each side, towards the east and west ends, alone remain. The transepts are in similar style, but have eastern chapels, and part of the Transitional aisled choir remains. The north and west sides of the central tower are also Transitional Norman. The western towers and

York Minster and Bootham Bar

west front are Early English. Of the Decorated period
are the east end of the choir, the lady loft or chapel (now
library), built over the chapter-house, and the sedilia
in the choir. The southern and eastern sides of the
central tower collapsed in 1458, and were rebuilt in the
Perpendicular style. Early in the sixteenth century
the nave, with the exception of the east and west ends,
was also rebuilt in this style, aisles being added. All
the towers had originally wooden spires covered with
lead, but the central one, having been injured by
lightning in 1593, fell through the roof in 1660, and
four years later the western spires were removed to
avert a similar catastrophe. The result is that the
towers, though 110 feet high, look low and stunted.

21. Architecture—(b) Military

Many of the earlier Norman castles in England con-
sisted of a flat-topped mound of earth, with a circular
palisade of timber around the summit, the whole surrounded by a dry ditch. These moated mounds were
called *mottes*, and examples occur at Castle Haugh,
Gisburn,[1] and Castle Hall Hill, Mirfield. Often there
was attached a bailey or base court of crescentic shape
on one side of the mound, as at Castle Hill, Almondbury; Castle Hill, Mexborough;[2] and Laughton-en-le-
Morthen.[3] William the Conqueror built two castles at
York. The first was situated on the tongue of land

---

[1] It is not absolutely certain that these are Norman.
[2] *Ibid.*   [3] *Ibid.*

Ripon Cathedral

between the Ouse and the Foss, where the existing castle stands ; the second was on Baile Hill on the opposite (right) bank of the Ouse, and was completed in eight days.   The chief existing remains of the ancient castle are 120 yards of the original curtain wall, two three-quarter round towers, and the keep, known as Clifford's Tower, of quatrefoil plan, built 1245 to 1259, when the wooden castle was replaced by a stone structure.

When the palisade of an early Norman castle was replaced by a wall of stone, we have a " shell keep," of which Clifford's Tower is an instance.   A better example is Tickhill Castle, where the foundations only of the decagonal keep remain on the top of a mound, about 60 feet high, and the bailey or ward has a curtain wall and an early Norman gatehouse.   The keep of Conisbrough Castle, though not a shell keep, is the most interesting and best preserved keep in the riding.   It is probably of the latter part of the twelfth century, and is circular, measuring 66 feet in external diameter at the ground level.   At equal intervals are six great buttresses, each projecting 9 feet more.   The keep is divided into five stages, the lowest a vault 22 feet in diameter, with a well in the centre.   Above it was the first floor, a store-room, with walls 15 feet thick, and at this level, 20 feet above the ground, was the entrance.   A staircase in the thickness of the wall leads to the second and third floors, in each of which there is a handsome fireplace.   On the third floor is an oratory or chapel.   The top stage was in the roof.

Pontefract Castle, of which a great part is Norman and part of fourteenth century, had a shell keep, usually described as of trefoil plan, but if a drawing in the Duchy of Lancaster Records may be trusted, it had at least six towers and eight bartizans or corbelled-out turrets. The inner ward had about seven other rectangular towers, and a detached (Swillington) tower ; it contains the foundations of a Norman chapel. In 1400 Richard II met his death in the castle, which between 1644 and 1649, underwent three sieges by Parliamentary forces, and after the last siege, was demolished by order of parliament.

Knaresborough Castle occupied about 2½ acres at the top of a cliff of magnesian limestone, some 230 feet above the Nidd. Its most interesting feature is the rectangular keep of the fourteenth century. In the third storey is the king's chamber, which is remarkable for the fact that the only communication between the inner and outer ward of the castle was through it. Of Sandal Castle, a mile south of Wakefield, very little remains. It covered nearly 6 acres, and had a circular keep, with three laterally projecting towers.

Nothing is left of Sheffield Castle, which was in existence in 1187, burnt in 1266, rebuilt 1270, and demolished by order of parliament in 1648 ; and of the Norman castle of Skipton, built by Robert de Romille in the time of the Conqueror, only the western doorway of the inner castle exists. About 1311 Robert de Clifford built the greater part of the existing castle, consisting of seven massive round towers, connected by apartments, and enclosing a large outer and an inner quadrangle. Henry de Clifford, created Earl of Cumberland by Henry VIII, built the more modern portion of the castle, and Lady Anne Clifford repaired the castle after its siege and surrender to the Parliamentary army in 1645. Harewood Castle, quite unlike those hitherto mentioned, was a compact quadrangular building without interior courtyard, built about 1367. Its chief feature is the great hall with a remarkable canopied Decorated recess, serving as a sideboard.

The walls of the city of York are 2 miles and 15 yards in length, mostly built on the top of an artificial bank of earthwork. They are of white magnesian limestone, have battlements, numerous bastions, and still more numerous buttresses. They were pierced by eight posterns, and five great gates or "bars," the older part of which is Norman. These bars—Bootham on the N.W., Monksbar on the N.E., Walmgate on the S.E., Fishergate on the S., and Micklegate on the S.W.—had originally no portcullises, but in the Decorated period these were added, and still remain in Bootham, Monk, and Walmgate Bars. The gatehouse had also added to it a superstructure with turrets or bartizans. In front of each gatehouse was a rectangular barbican or enclosure, with battlements and an outer gate. This has been restored at Walmgate.

Skipton Castle

22. Architecture—(c) Domestic

It is not always easy to draw a sharp line between military and domestic architecture, for some buildings partake of both characters.

In Anglo-Saxon times dwellings were everywhere of wood, indeed, until the close of the sixteenth century many of the houses of the less wealthy were wooden. Leland, in his " Itinerary " (1535–1543) says, " The hole toune of Dancaster is buildid of wodde." Few timber or half-timbered houses have been preserved in the West Riding, but among noteworthy examples of the latter are Shibden Hall near Halifax; "Six Chimneys," Kirkgate, Wakefield; Wormalds Hall, Almondbury; Lees Hall, Thornhill; "The Old Nookin," at Oulton near Leeds; the "Bishop's House," Meersbrook Park, Sheffield; and in York, a timber house at the end of the Pavement, and a house in Newgate, near the Shambles.

We may turn now to the more important houses built of stone or brick, taking them in chronological order.

The ruins of Spofforth Castle, between Harrogate and Wetherby, formed originally one side of a quadrangle of which the rest has disappeared. More than half the length is occupied by the great hall, the basement of which is Trans. Norman, the rest Decorated but probably rebuilt in the 15th century. The royal licence to Henry de Percy, to crenellate (fortify) is dated 1308. Most of the rooms have large pointed windows, with remains of Geometrical tracery. From one angle there projects an octagonal tower, containing a newel staircase, and capped by a pyramidal roof.

John de Merkingfield obtained a license to crenellate Markenfield Hall, 3 miles S.S.W. of Ripon, in 1310. In plan the house is L-shaped, and is crowned by an embattled parapet. The great hall has four large Geometrical windows of two lights, and the chapel one of three lights ; the rest of the windows are square-headed. There is a tower with newel stair, and its original pyramidal roof. The architecture generally resembles that of Spofforth Castle.

Bolton (by Bowland) Hall is beautifully situated near the Ribble. Though much altered in modern times, it contains a great hall with open timbered roof, a panelled chamber with square-headed mullioned window, and a chamber above it called Paradise, all possibly of the reign of Edward III, and interesting from the fact that it was here that Henry VI took refuge after the battle of Hexham in 1464.

The Archbishops of York possessed a manor house at Cawood before the Conquest, but licence to crenellate was not obtained until 1271. Of their Cawood Castle, the only remains are the Perpendicular gatehouse, built between 1426 and 1451, with a beautiful oriel window, and an adjoining red-brick chapel, now a barn. Here Cardinal Wolsey was arrested on his downfall in 1530.

Barden Tower, 2½ miles N.W. of Bolton Abbey, is in ruins. It was built by Henry Clifford, the " Shepherd Lord," in the latter part of the fifteenth century, and was restored by Lady Anne Clifford in 1658.

Bolling (Bowling) Hall, 1 mile S. of Bradford, is now a municipal museum. This once stately mansion is chiefly Elizabethan, though parts are older. It contains a fine oak chimney-piece, and a good example of ornamental plaster ceiling.

Woodsome Hall, 2½ miles S.E. of Huddersfield, is partly of the time of Henry VIII, with additions in 1600 and 1644. It is a most charming house in the Elizabethan style, built round an oblong open courtyard.

Heath Old Hall, 1 mile E. of Wakefield, was built by John Kaye in the reign of Elizabeth. It is of three storeys and nearly square in plan. In front and at the angles the bay windows are carried up to form octagonal turrets. At the sides are oriels, and the remaining windows are square-headed, stone mullioned, and of many lights.

Weston Hall, 2 miles N.W. of Otley, built by the Vavasours in the reign of Elizabeth, has semicircular-headed window lights, and in the gabled wing a large bay window of four storeys.

Swinsty Hall, 5 miles N. of Otley, built in 1560 or 1570, is Elizabethan in architecture. It is oblong in plan, with two projecting wings in front, four gables, and square-headed stone mullioned windows of many (up to ten) lights.

The existing long, three-storied façade of Browsholme Hall, 4¼ miles N.W. of Clitheroe, dates from 1604, though parts of the hall are of the sixteenth century.

Woodsome Hall, near Huddersfield

The architecture is Jacobean Renaissance.   The main doorway has on each side coupled columns, Doric, Ionic, and either Corinthian or Composite in three storeys.

The most beautiful old hall in the riding is perhaps

Fountains Hall

Fountains Hall.   It was built in the Jacobean style by Sir Stephen Proctor in 1611, with stones taken from the neighbouring Fountains Abbey.   The façade shows a central doorway with coupled Ionic columns, while at each end is a rectangular tower.

Temple Newsam, 2 miles E.S.E. of Leeds, was built by Sir A. Ingram in 1630. It is of brick, with stone coigns, and occupies three sides of a quadrangle. It has no gables, but has many large bay windows, and is surmounted by a parapet of letters cut in stone, forming a long pious inscription.

Wentworth Woodhouse, 3½ miles N.W. of Rotherham, is the seat of Earl Fitzwilliam. It has a facade 600 feet long in the Italian Renaissance style, with a fine portico in the centre. It incorporates large portions of the older, probably Tudor, mansion, once the home of the great Lord Strafford. The garden front was designed by Delin, the park front by Henry Flitcroft in 1737.

Harewood House, 7 miles N. of Leeds, was built in 1760 by John Carr, a well-known York architect. It is in the Italian Renaissance style, in plan a parallelogram of about 247 by 84 feet with Corinthian columns and pilasters. Additions, designed by Barry, were made in 1843.

23. Communications—Past and Present—Roads, Canals, Railways

The earliest roads in Britain were mere trackways connecting hamlet with hamlet, and owing to the marshy nature of the low-lying ground, and the thick growth of underwood, they usually kept along the drier heights. The excellent system of Roman roads and their construction have already been described in Chapter

18.   After the Romans left, they were adopted by their successors in later times. The repair of roads and bridges was part of the *trinoda necessitas* or triple obligation of landowners of Anglo-Saxon times, but was often neglected. Richard de Kellawe, Bishop of Durham (1311–1316), remitted forty days penances to those who helped in the building or maintenance of the causeway between Brotherton and Ferrybridge. Other means of maintaining roads and bridges in the Middle Ages were special gilds, endowments, tolls, and the pious offerings at the chapels on the bridges. These chapels still exist at Wakefield and Rotherham, and that on the old Ouse Bridge at York, was only demolished in 1810. Macaulay graphically depicts the badness of the English roads in 1685. In 1663 the first Turnpike Act was passed, authorising the Justices of Hertford, Cambridge, and Huntingdon to collect tolls to be applied to putting a particular road in repair. In 1740 an Act was passed to repair and enlarge the roads from Leeds to Selby, and Leeds (by two routes) to Halifax. Between 1740 and 1759–60 six other Acts were passed, dealing with turnpikes (as these roads subject to toll were called) from Leeds ; that of the last-mentioned year was for improving the road from Leeds through Wakefield to Sheffield, which later became one of the great coach roads to London, the other being through Doncaster.

These Turnpike Acts were very unpopular with the users of roads, who; from time immemorial, had had free use of the road, though it was often almost impassable. On June 18th, 1753, mobs from Otley and

Yeadon joined forces, and in a week destroyed about a dozen turnpike gates and houses. At Selby still more damage was done. In the same month at Leeds three men were imprisoned for refusing to pay toll, and a riot ensuing, eight rioters were shot dead, and more than forty wounded. The determination to get rid of the Turnpike Trusts and their tolls dates from the 'sixties of the last century. In 1871 there were still 854 left, in 1883 only 71.

About the year 1600 long lumbering broad-wheeled vehicles, called stage-waggons, drawn by eight or ten horses, were employed, but until about the middle of the seventeenth century, the ordinary mode of travel was on foot or on horseback. Stage coaches first came into use in England in 1640, and in 1658 one ran thrice a week from London to York, in four days, for eleven shillings. In 1836 the same journey, of 197 miles, was made by coach in twenty hours, including stoppages. In 1695 Sir Walter Calverly mentions a coach which ran regularly from Wakefield through Barnsley and Sheffield to London. Defoe describes how, about 1727, the whole-sale cloth merchants of Leeds went all over England with droves of packhorses, laden with cloth, and it was quite an ordinary thing for one of them to carry cloth to the value of £1000 with him. In 1760 " flying machines on steel springs " were advertised to run from Leeds, Wakefield, Barnsley, and Sheffield to London.

The great coaching era began in 1786, when the Royal Mail was first carried by coach, and ended in 1842 when the Royal Mail coach, which had run regularly from

London to York since 1786, finally went off the road. In 1838 there were 65 coaches leaving and 65 arriving at Leeds daily.

The most celebrated road in the riding is the Great North Road from London. It enters Yorkshire at Bawtry, and the nine miles thence to Doncaster (partly Roman road) form one of the finest stretches of road in England. From Doncaster to Ferrybridge, 15 miles, the road—which partly coincides with Ermine Street— is very hilly. At Brotherton the Great North Road bifurcates. The western branch joins the Roman Road through Aberford, follows it for 2 miles, and then goes through Wetherby to Boroughbridge, where it leaves the West Riding for Catterick, Bowes, and Appleby. The eastern branch, which runs through Sherburn-in-Elmet to Tadcaster, has scarcely a level bit in it, but thence to York ($9\frac{1}{2}$ miles) is almost a dead level. At York the Great North Road leaves the West Riding and traverses the North Riding by Easingwold, Thirsk, and Northallerton.

The cost of transporting goods by road was so heavy that canals came into use. The Company of the " Undertakers of the Navigation of the Rivers Aire and Calder in the West Riding of the County of York," was established by Special Act in 1698, and reduced the cost of transporting heavy goods to $\frac{1}{4}$th and even $\frac{1}{10}$th of the rates before prevailing. The main trunk line runs from Goole to Castleford, where it divides into two branches, one on the line of the Aire to Leeds, where it joins the Leeds and Liverpool Canal, the other on the line of the

Calder to Wakefield, where it joins the Calder and Hebble Navigation. The Barnsley Canal, 14 miles long, opened in 1799, is part of the system, and runs from Wakefield to Barnsley, where it joins the Sheffield and South Yorkshire Navigation. The latter company is joint owner of the New Junction Canal, 5¾ miles long, from Bramwith to Sykehouse, which connects the two systems.

The " Company of Proprietors of the Canal Navigation from Leeds to Liverpool " was incorporated as such in 1770, but some parts of the canal were not completed until 1816. The main line is 127 miles long (of which about 40 miles lie within the riding) and there are 99 locks. From Leeds it follows the line of the Aire by Shipley, Bingley, and Skipton to Gargrave, and thence by Barnoldswick to Nelson, Burnley, Accrington, Blackburn, Chorley, and Wigan to Liverpool.

The Calder and Hebble Navigation Company, established 1757, has a main line opened about 1780 from Wakefield to Sowerby Bridge, where it connects with the Rochdale Canal : it has a branch to Halifax, completed in 1829. The " Company of Proprietors of the Rochdale Canal " came into existence in 1794 ; and its waterway follows the line of the Calder from Sowerby Bridge to Todmorden, whence it runs by Rochdale and Failsworth to Manchester. Its total length is 32 miles, of which about 12 are in the riding. It has 92 locks and two tunnels.

The Huddersfield Canal joins the Calder and Hebble Navigation and follows the Colne through Huddersfield to Marsden, where there is a tunnel nearly 3½ miles

Five-Rise Locks, Bingley

long. It then accompanies the Tame by Stalybridge, and joins the Ashton Canal near Manchester. About 17 miles of its length are within the riding. In the 20 miles of narrow canal there are 74 locks.

The " Company of Proprietors of the Navigation of the River Don " originated as far back as 1732 and opened up navigation from Fishlake to Tinsley, and in 1815 a company obtained powers to make a canal from Tinsley to Sheffield. In 1793 a canal was authorised from Stainforth on the Don to Keadby on the Trent, and another—the Dearne and Dove Canal—from Swinton to Barnsley. This was opened in 1804. All these four undertakings were afterwards acquired by the Sheffield and South Yorkshire Navigation Co., incorporated in 1889.

The Ouse is navigable from the Humber estuary to the confluence of the Swale and Ure, and the Ure is canalised up to Ripon under the name of Boroughbridge and Ripon Canal.

The following table shows the total length of each of the canal systems and the tonnage of goods transported in 1913 :

	Miles.	Tonnage.
Aire and Calder Navigation . .	85	3,597,921
Leeds and Liverpool Canal Co. . .	145	2,308,210
Calder and Hebble Navigation . .	27	482,983
Rochdale Canal Co. . . .	34	512,061
Huddersfield Canal . . .	24	91,753
Sheffield and South Yorkshire Navigation Co.	60	920,876
Boroughbridge and Ripon Canal .	10	9,437

Turning now to railways and beginning with the lines constituting the North-Eastern Railway, the main line of the East Coast route from London to Scotland traverses the riding by the comparatively level Vale of York, from Shaftholme Junction, north of Doncaster, where it joins the Great Northern, to Selby, 14 miles (1871 [1]) and continues through the East Riding to York and the North Riding to Northallerton, etc. The Leeds and Selby line, 19¾ miles (1834) is the oldest in the riding. Other North-Eastern branches are Church Fenton to Spofforth (1847) continued to Harrogate in 1848, in all 18½ miles ; Poppleton Junction, near York, to Knaresborough, 14½ miles (1848–51) ; Leeds to Ripon, 29¼ miles (1849) ; Harrogate to Pateley Bridge, 11½ miles (1862) ; Arthington to Ilkley, partly joint line (1865) ; Knaresborough to Boroughbridge, 7¼ miles (1875). The Midland and North-Eastern Swinton and Knottingley joint line, 16¾ miles (1879), in conjunction with the Knottingley to Burton Salmon N.E. branch and Burton Salmon to York N.E. line, affords a direct route by Midland from York to Bournemouth.

The main line of the Great Northern Railway, forming part of the East Coast route from London to Scotland, enters the riding at Bawtry, proceeding to Doncaster, 8¼ miles (1849), thence to Shaftholme Junction, where it joins the North-Eastern, as before mentioned. From Marshgate Junction, near Doncaster, a Great Northern Railway line runs to Wakefield, continuing to Holbeck

[1] The dates within brackets are those of the opening of the respective lines.

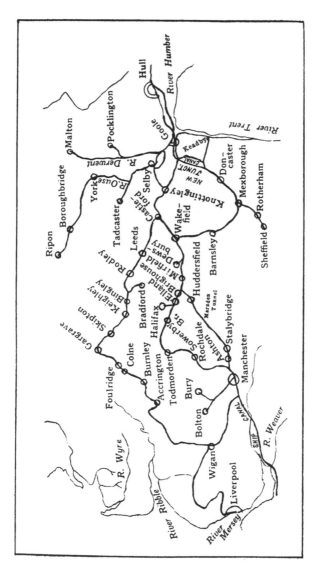

Canal System of the W. Riding and its Connections

and Leeds. From Leeds the Great Northern runs through Holbeck to Bradford (1854) and continues by Queensbury to Thornton and Keighley. From Queensbury a branch runs to Holmfield and by Lancashire and Yorkshire joint line to Halifax. From Wakefield there is a Great Northern line through Ossett to (Upper) Batley, and by Drighlington to Bradford. There are branches to Shipley, Low Moor, Methley, etc.

The oldest portion in the riding of what is now the Midland Railway is the line from Sheffield to Rotherham, 5½ miles, constructed in 1838. The Midland main line from London to Carlisle and Scotland traverses the riding as follows—the line from Derby to Leeds built in 1840 enters the riding south of Woodhouse Mill station and passes through Masbrough (Rotherham), Swinton, and Normanton ; Leeds was connected with Shipley (10¾ miles) in 1846 ; Shipley with Skipton, (15¼ miles) in the following year. The line from Skipton to the junction for Settle, continued by Clapham and Bentham to Morecambe, dates from 1849 ; but that from Settle to Hawes Junction and Carlisle was only built in 1875. From Methley through Leeds and Skipton to Bell Busk the line follows the valley of the Aire, making use of the Aire Gap to reach the valley of the Ribble, which it follows from Hellifield to Ribblehead station. Branches run from Wincobank (near Sheffield) to Barnsley, Royston to Dewsbury, and Skipton to Ilkley, and there are six other branches. Realising the mistake of leaving Sheffield on a branch line the company connected it by a new line with Chesterfield in 1870.

The London and North-Western Railway affords communication between Manchester and Leeds, running over the Lancashire and Yorkshire line as far as Stalybridge, and entering Yorkshire a mile S.W. of Greenfield station. At Diggle it reaches the Standedge tunnel (3 miles 62 yards) under the Pennine water-parting, and follows the Colne valley to Huddersfield and Heaton Lodge junction, between which and Dewsbury junction the trains run down the Calder valley over the Lancashire and Yorkshire line, thence viâ Dewsbury, Batley, and Morley to Leeds Central, or from Farnley and Wortley to Leeds New Station. From Heckmondwike a line runs by Liversedge and Cleckheaton to a junction with the main line, south of Farnley and Wortley station. A branch of the London and North-Western Railway joins a Midland branch line at Ingleton and leads northward to Sedbergh and Carlisle.

The Lancashire and Yorkshire Railway in the West Riding is chiefly a west to east line, extending from Manchester on the south and Burnley on the north to Goole. The line from Manchester enters the riding in Littleborough tunnel, reaches the Calder valley at Todmorden, where it is joined by the line from Burnley, which runs through the Calder Gap. The railway follows the Calder valley downward through Wakefield to Normanton (opened from Manchester 1841) where it joins the North-Eastern line to York. From Wakefield the line runs to Pontefract, Knottingley, and Goole. From near Sowerby Bridge a line runs through Halifax and Low Moor to Bradford. It is joined at Low Moor

K

by the branch from Mirfield, through Cleckheaton, constructed in 1848. A branch, opened 1850, runs from Huddersfield to join the Great Central at Penistone, another from Horbury to Barnsley, a third from Knottingley to Askern junction where it joins the Great Northern from Doncaster. The Lancashire and Yorkshire line from Manchester and Blackburn enters the riding S.W. of Rimington station and joining the Midland at Hellifield provides a route to and from Scotland.

Although the Great Central Railway has had a main line from London since 1899, it is essentially a west to east line in the riding. The main line enters the riding a mile S.E. of Woodhouse station, where it is joined by the line from Grimsby and Retford. Thence it continues for 5 miles to Sheffield. The main line from Sheffield to Manchester, opened in 1845, follows the valley of the Don to Dunford Bridge, where it enters the Woodhead tunnel (3 miles 17 yards) under the Pennine water-parting. From Sheffield the Great Central runs north-eastwards following the valley of the Don to Rotherham, Mexborough, Doncaster, and Thorne, and by Keadby to Grimsby. At Thorne (Waterside) it joins the North-Eastern Railway to Goole. Branches run from Swinton by Barnsley and Nostell junction on the Great Northern Railway to Wakefield; from Meadowhall, near Wincobank, to Chapeltown; and from Barnsley to Penistone.

The Hull and Barnsley Railway runs from Hull to a junction with the Midland Railway at Cudworth near Barnsley, and has a branch from Wrangbrook junction to Wath, made in 1902. The main line crosses the Ouse between Barmby and Drax stations.

The number of men employed on railways in the West Riding in 1911 was 34,885, besides 3095 in York, and exclusive of 5826 railway coach and waggon makers.

24. Administration and Divisions– Ancient and Modern.

The division of Yorkshire into ridings, which are so arranged that all three meet at York, probably occurred in the latter half of the ninth century, when Danish kings reigned at York. The ridings doubtless correspond to a certain extent to older divisions; the West Riding, for instance, to the British kingdom of Elmet. Instead of speaking of division we ought rather, perhaps, to use the word aggregation. " The hundred," writes Professor Freeman, " is in truth formed by an aggregation of marks [roughly represented by the modern parish or manor], the shire by an aggregation of hundreds, the kingdom by an aggregation of shires." The making of the riding into wapentakes, which correspond to the hundreds of southern counties, probably dates from the Danish period just mentioned. The word " wapentake," derived from the Norse, means literally " weapon taking " or " weapon touching," and was applied to the form of ratifying the decisions of the local court. The wapentakes of the West Riding as recorded in Domesday Book (1086) were as follows :—

Ancient Wapentake.	Geographical Situation.	Modern Wapentake.[1]
Strafordes . . .	S. of the upper and lower Don, and both sides of middle Don.	Strafforth and Tickhill.
Staincros or Stancros .	Both sides of upper Don and Dearne.	Staincross.
Hagebrige or Agebruge .	S. of the upper, and both sides of lower Calder.	Agbrigg and Morley.
Morelei or Moreleia .	Chiefly between Aire and Calder.	
Osgotcros . . .	Between lower Don and Aire.	Osgoldcross.
Siraches . . .	Between middle Aire and Wharfe.	Skyrack.
Barcheston or Barchestone .	Between lower Aire and Wharfe.	Barkston Ash.
Annesti or Einesti .	Between lower Wharfe and Nidd.	Ainsty.
Borchescire or Borgescire .	Between middle Wharfe and lower Ure.	Claro.
Halichelde . . .	Between lower are and Swale.	Hallikeld in North Riding.
Crave or Cravescire, a district, not called a wapentake .	Upper parts of basins of Wharfe, Aire, and Ribble.	Staincliffe and Ewecross (Craven District).

[1] The Wapentakes are now only used as rating areas for the repair of bridges.

In Anglo-Saxon times each township had its town reeve, who with its four best men represented it at the Wapentake Court, held every month, and at the Shire Moot. At the Wapentake Court there were two principal officials, the reeve of the wapentake (bailiff after the Conquest) and the *hundredes-ealdor*, who convened and was the constituting functionary of the court. The wapentake formed a rateable division of the county. The Shire Moot (which was not only a court, but an assembly of the people) was held twice a year—that is the full court—but according to the charter of 1217 it met once a month. Its constituting official was the shire-reeve or sheriff, who was accompanied by the ealdorman (called earl under the Danish kings) or governor of the shire, and the bishop, who respectively declared the secular and spiritual law. The ealdorman often governed more than one shire. He received one third of the profits of the court and commanded the military force of the whole district under his sway, becoming sometimes so powerful as to be a danger to the Crown. William the Conqueror realised this after the conspiracy of the earls in 1074 and appointed few earls, relying on the sheriffs. The sheriff in Anglo-Saxon times was " the king's steward and judicial president of the shire, the administrator of the royal demesne, and executor of the law " ; in Norman times he was also the king's military representative. The older sources of income, that is those available under the Saxon kings, were farmed, at any rate under the Norman kings, by the sheriff, a system necessarily leading to extortion.

In 1340 it was provided by statute that no sheriff should continue in office for more than a year. In the eighteenth and nineteenth century the practical power of the sheriff was further diminished. Even by 1689 he discharged his functions by deputy, appointing a professional Under Sheriff, usually the Clerk of the Peace and the same person who had served his predecessor in office. The chief duty of the sheriff at present is to receive the judges of assize with ceremony. There is only one High Sheriff for Yorkshire..

The office of Lord Lieutenant is a life appointment and dates only from the middle of the sixteenth century ; from the reign of William and Mary it was usually combined with that of *Custos Rotulorum* or " Keeper of the Rolls of the Peace." The Lord Lieutenant appointed the officers of the militia and took command in an emergency of all the local forces of the county. The Justices of the Peace are appointed upon his nomination and he is president of the Territorial Association of his county. During the eighteenth and nineteenth centuries the local government of counties fell largely into the hands of the Justices of the Peace in Quarter Sessions. By the Local Government Act of 1888 County Councils were established and took over the functions of local government, previously fulfilled by the justices, with additional functions. Each riding of Yorkshire is a separate administrative county, with a separate county council. The West Riding county council consists of 90 councillors, elected in 90 electoral divisions, each returning one councillor. They hold office for 3 years. There are, in addition, 30 aldermen, who are elected by the councillors and hold office for 6 years. By the Act of 1888 boroughs were divided into three classes: (1) The County Borough, *i.e.* boroughs already counties of themselves or having population of not less than 50,000. They are completely independent of the County Council, but have all its powers. The county boroughs of the West Riding are the cities of York, Bradford, Leeds, Sheffield, and Wakefield, and the boroughs of Barnsley, Dewsbury, Halifax, Huddersfield, and Rotherham. (2) The larger Quarter Sessions Boroughs, with a population exceeding 10,000. These are subject to the control of the County Council for certain special purposes only. (3) The boroughs with a population of under 10,000. Subordinate to the County Council there are in the West Riding 112 Urban District Councils (exclusive of the boroughs) and 28 Rural District Councils. By the Local Government Act of 1894 Parish Councils were established, of which there are 256 in the West Riding. There are 650 townships or civil parishes in the riding not including those in the county boroughs.

The parliamentary representation of the West Riding is as follows: Parliamentary boroughs, Sheffield returning 7 and Leeds 6 members, Bradford 4, and Barnsley, Batley and Morley, Dewsbury, Halifax, Huddersfield, Rotherham, and Wakefield 1 each. There are 19 County Divisions, each returning 1 member: Barkston Ash, Colne Valley, Doncaster, Don Valley, Elland, Hemsworth, Keighley, Normanton, Penistone, Pontefract, Pudsey and Otley, Ripon, Rother Valley,

Rothwell, Shipley, Skipton, Sowerby, Spen Valley, and Wentworth.

Ecclesiastically the riding contains the whole of the dioceses of Wakefield, Sheffield, and Bradford, the southern part of the diocese of Ripon, and a small part of that of York.

25. The Roll of Honour

The West Riding has been the birthplace or the home of many distinguished men of whom only a few can be noted here.

Of soldiers and statesmen Ferdinando, 2nd Baron Fairfax of Cameron, Parliamentary General (born 1584, died 1648) and his son Thomas, the great Lord Fairfax, born at. Denton 1612, died 1671, and William Cavendish, Duke (Earl) of Newcastle who was born at Handsworth near Sheffield in 1592, have already been noticed in the historical section.   Thomas Wentworth, the great Earl of Strafford, was born in London in 1593, but his family had been long established at Wentworth Woodhouse, near Rotherham.   He was one of the chief advisers of Charles I, was condemned by a Parliamentary bill of attainder, and was beheaded in 1641.   Thomas Osborne, 1st Duke of Leeds (in Kent) better known as Earl of Danby, was Lord High Treasurer to Charles II, and was born probably at Kiveton in 1631.   Sir William Gascoigne, who was born at Gawthorpe about 1350 and became Chief Justice of the King's Bench, is immortalised by Shakespeare in connection with the story of his having committed Henry V to prison when Prince of Wales for having assaulted him when on the Bench.

Sir Martin Frobisher, one of the most distinguished of the Elizabethan seamen was born about 1535 at Altofts. He made three voyages to the Arctic regions originally

Thomas, 3rd Baron Fairfax, and Lady Fairfax

in search of a North-West passage, and was knighted for his valour against the Armada.

Another Armada hero, scholar and great favourite of Queen Elizabeth, was George Clifford, 3rd Earl of Cumberland, who, though not born in the riding, was the owner of Skipton Castle.

The riding is rich in distinguished theologians and divines. Thomas Rotherham, who lived to become Archbishop of York and Lord Chancellor of England, was born at Rotherham in 1423. He was educated at King's College, Cambridge, of which he became a considerable benefactor. To Lincoln College, Oxford, his benefactions were even greater, adding both to its buildings and its revenues. He died in 1500.

John Tillotson, who was born at Old Haugh End, Halifax, in 1630, was perhaps the most popular preacher that England has known, and his sermons were regarded as "the standard of finished oratory." At first a Calvinist, he submitted to the Act of Uniformity at the Restoration and married a niece of Oliver Cromwell two years later. He rose rapidly in office, became Dean of St Paul's, succeeded Archbishop Sancroft in the see of Canterbury in 1691, and died in 1694.

The parents and forebears of William Paley, the moral philosopher (born 1743), whose "Evidences of Christianity" formed until 1920 a subject of examination at Cambridge, lived at Giggleswick, at which place he spent his youth. Entering at Christ's College he became Senior Wrangler and ultimately fellow and tutor, and resided ten years in the University, dying as Sub-Dean of Lincoln and rector of Bishop Wearmouth in 1805. Henry Venn (born at Barnes in Surrey 1725, but Vicar of Huddersfield 1759-71, died 1797), one of the best known of the Evangelical leaders, is perhaps even more worthy of fame as founder of the Church Missionary Society ; and Joseph Bingham's great work on

Joseph Priestley

the History of Christianity—*Origines Ecclesiasticæ*—
gives him high rank among the theologians of his
time. He was born at Wakefield in 1668 and died
in 1723.

Richard Bentley, the great classical scholar, was
born at Oulton in 1662 and after many preferments
ultimately became Master of Trinity College, Cambridge.
He had a stormy reign, being a born fighter, but he
managed to find time to publish a large number of
volumes which reveal remarkable scholarship. He
died in 1742.

In the domain of science the riding is perhaps not
so distinguished as in letters, but it can show some well-
known names. Joseph Priestley, the chemist, who was
born at Fieldhead near Birstall in 1733, was a Wesleyan
minister. He discovered oxygen and other gases and
the composition of water, wrote many religious books,
and died in America in 1804. Charles Waterton (1782–
1865), less the serious naturalist than the nature-loving
traveller, was long in South America and other parts
of the New World, and made his ancestral home at Walton
Hall near Wakefield a sanctuary for birds and beasts.
The riding has produced several geologists : John Phillips
(1800–1874), though not a Yorkshireman, was for many
years Curator of the Museum at York, and published
Illustrations of the Geology of Yorkshire; Henry Clifton
Sorby, who was born at Sheffield in 1826 and died in
1908, has been called the Father of Microscopic Petro-
graphy; and Adam Sedgwick, Woodwardian Professor
of Geology at Cambridge, born at Dent in 1785, died

Adam Sedgwick

1873, was one of the most distinguished geologists of his day.

To Henry Briggs, who was born at Warley near Halifax in 1556, the science of mathematics is greatly indebted, and still more the seaman, for he was the author of *Arithmetica Logarithmica* and as the introducer of the use of logarithms he has been called the greatest bene-factor the Navy ever had.   Educated at Cambridge he became Savilian Professor of Astronomy at Oxford, and was a patron of the Arctic voyage of Luke Foxe, a fellow-Yorkshireman, who named a group of islands after him " Briggs his Mathematicks " in 1631, the year following his death.   The name of John Radcliffe, the physician, is known to every dweller in Oxford, and is likely to remain so for all time, for he left great wealth to the University, of which the Radcliffe Library, Infirmary and Observatory are the visible signs.   He was born in 1650 at Wakefield and died in 1714.

Though the name of John Harrison is little known, the world owes him a great debt.   He was the inventor of the chronometer and various improvements in time-pieces and was born at Foulby in 1693.   John Smeaton (born Austhorpe 1724), built in 1755 the great Eddystone lighthouse, the third in number, which stood till 1882.   By most, Joseph Bramah (1748–1814) is chiefly remembered as the constructor of an ingenious lock, but he was an inventor of great ability, who has the hydrostatic press, the beer-drawing machine, and the idea of the screw propeller to his credit.   He was born at Stainborough near Barnsley.

Numerous antiquaries have been natives of the riding. Ralph Thoresby (born at Leeds 1658, died 1725) wrote two books on Leeds—the *Ducatus Leodiensis* and *Vicaria Leodiensis.* His contemporary, Abraham de la Pryme, (1672–1704) though also an antiquary, is best known as a diarist. Of rather later times are John Burton (born at Ripon 1697, died 1771), physician and author of *Monasticon Eboracense,* and Francis Drake (born at Pontefract in 1696, died at Beverley 1771) who practised medicine in York, and wrote *Eboracum,* a history of that city. Joseph Hunter, a Sheffield man (1783–1861), by profession a Presbyterian minister, was author of several books on Yorkshire, notably on the history and topography of Hallamshire, Sheffield, and South Yorkshire. Among the earlier archæologists, John Potter (born at Wakefield 1674, died 1747) was famous : he wrote *Archaeologia Graeca,* which enjoyed extraordinary popularity and passed through many editions. He was made Archbishop of Canterbury in 1737.

One of the most learned scholars of his age was Sir Henry Savile (born at Bradley 1549, died 1622), tutor to Queen Elizabeth, Warden of Merton College, Oxford, author of many books, and founder of the Savilian professorships of geometry and astronomy at that University. In modern days another Oxonian of note appears as a native of the riding—William Stubbs (born at Knaresborough in 1825, died 1901) Regius Professor of Modern History in Oxford, and afterwards Bishop of Chester (1884) and of Oxford (1889). His works on history are numerous ; conspicuous among them being his admirable *Constitutional History of England*.

At the head of the lengthy list of writers, whether of plays, poetry, or fiction must be placed William Congreve

William Congreve

(born Bardsey 1669, died 1729) one of the greatest of English comedy writers. He was a master of brilliant repartee, and his *Love for Love* has been termed the finest prose comedy in the English language. Edward Fairfax, who was born at Leeds about 1580 and died in 1635, is noteworthy as the translator of Tasso's *Gerusalemme Liberata* into English in 1600. Although by birth a Scotsman, James Montgomery (1771–1854) spent so much of his life in the riding, and was so intimately connected with its political history that his name cannot be omitted. To most people he is only known as a writer of hymns, many of which still hold their place in our hymnals, but he began life as a Radical newspaper editor, and started the *Sheffield Iris*. Alfred Austin, who succeeded Tennyson as Poet Laureate in 1896, was born at Headingley in 1835 and died in 1913.

The Brontë sisters, a remarkable trio of authoresses, whose father's name was originally Prunty, were all born at Thornton near Bradford; Charlotte (1816–55), Emily (1818–48), and Anne (1820–49). Their literary career began with a volume of poems under the pseudonyms of Currer, Ellis, and Acton Bell; and later each wrote novels. Charlotte still remains a name of mark in the world of letters, *Villette*, *Shirley*, and *Jane Eyre* being her best known books, but her sisters did not attain her eminence, though Emily's *Wuthering Heights* is of singular merit. Mrs Ewing, writer of many charming children's books (1842–85) was a native of Ecclesfield, and George Gissing (1857–1903), whose powerful but gloomy novels found many readers, was born at Wakefield.

Many artists of distinction have been connected with the riding. Son of a Leeds clothier, Benjamin Wilson (1721–88) earned fame as a portrait painter in London,

L

but was also a student of chemistry and electricity, an
F.R.S., the manager of a private theatre, and an en-
graver of merit. William Etty, R.A., born at York in
1787, the greatest master of flesh tints, has perhaps
scarcely received his full meed of praise even yet. He
died in 1849. Theodore Nathan Fielding, who lived at
Halifax, was an artist and father of four artist sons,
Theodore, Thales, Newton, and A. V. Copley Fielding,
all water-colourists, but the last named (1787–1855) by
far the most gifted,   Thomas Creswick, R.A., a pleasing
rather than masterly landscape painter, was born at
Sheffield in 1811 and died in London in 1869. William
Powell Frith, R.A., was born at Aldfield in 1819 and died
in 1900. His crowded, but well-grouped canvases of
the " Derby Day," " The Railway Station," etc., are still
familiar to-day. Among more modern painters are
Ernest Crofts, R.A. (1847-1911), and P. W. May (Phil
May), one of the most gifted of humorous draughtsmen,
born in Leeds in 1864, and died in 1903.   John Flaxman,
the sculptor (1755–1826) son of a modeller, was born in
York.   Apart from his more ambitious work, he became
well known to the public by his charming and graceful
classical figures reproduced in the Wedgwood pottery.

Among musicians may be mentioned Sir William
Sterndale Bennett, pianist and composer (Sheffield 1816–
75) ; John Curwen, founder of the Tonic Sol-fa system
of musical teaching, born at Heckmondwike in 1816 ;
J. T. Carrodus, violinist, born at Braithwaite in 1836 ;
and Sir Joseph Barnby, composer, who was born at
York in 1838.

26. The Chief Towns and Villages of the West Riding

(The figures in brackets give the population in 1911. The other figures are references to the pages in preceding chapters.) Abbreviations :—Sax. = Saxon, Norm.= Norman, E.E = Early English, Dec. = Decorated. Perp.=Perpendicular, Trans.=Transitional.

Adel cum Eccup (1083), 5 m. N.N.W. of Leeds. Roman station, p. 101. Norman church, p. 109.

Adlingfleet (179), near confluence of Ouse and Trent, has interesting E.E. church.

Adwick-le-Street (5723), colliery village, 4 m. N.N.W. of Doncaster, and 1 m. E. of the Roman Ermine Street. Church in part Norm., E.E., Dec., and Perp.

Aldborough (422), ¾ m. E.S.E. of Boroughbridge. Roman station, p. 101.

Almondbury (15,485), in county borough of Huddersfield. Church, p. 112. Earthworks, p. 124. Industries, p. 66.

Altofts (4689), colliery village, ¾ m. N.W. of Normanton Station. Experimental gallery for coal-dust explosions ; birthplace of Frobisher, p. 153.

Anston, North and South (2184), parish, 3 m. W.N.W. of Shireoaks. South Anston, church chiefly Dec., with a Trans. Norm. arcade and Perp. tower.

Ardsley (6870), urban district, 2 m. E. of Barnsley, has collieries and glass bottle works.

Ardsley, East (4441) **and West** (3679), urban district of two parishes, 4 m. N.W. of Wakefield. E. Ardsley church has Norm. S. doorway. There are blast furnaces and collieries here.

Arksey (468), village (eccl. par.), 2¼ m. N.N.E. of Doncaster. The church is partly Norm. and Trans. Norm.

Austerfield (370) 1½ m. N.N E. of Bawtry, has a Roman Camp. The Church has some Norm. work. Birthplace of William Bradford (1590-1657), one of the pilgrims of the *Mayflower*, and second Governor of Plymouth.

Baildon (6042), 1½ m. N. of Shipley, has worsted manufactories. Antiquities on Baildon Moor, p. 98.

Balby with Hexthorpe (11,570), 1½ m. S.W. of Doncaster, has tanneries, brickyards in boulder clay, p. 42. Here George Fox (1624–91) held the first meeting of the Society of Friends.

Barden Tower, ruins, 2½ m. N.W. of Bolton Abbey, p. 131.

Bardsey cum Rigton (302), 7 m. N.N.E. of Leeds. Church partly Sax. and Norm., pp. 109, 111. Birthplace of Congreve, p. 160.

Barnoldswick (9703), 4 m. N. of Colne, has cotton manufactories, p. 71. The chancel of the church of St Mary-le-Gill is E.E., the rest Perp.

Barnsley (50,614), is a county borough returning one M.P. It is on the River Dearne, and on the Midland, Great Central, and Lancashire and Yorkshire Railways, the Barnsley Canal, and the Dearne and Dove Canal. It is a coal mining centre, and has linen mills and glass bottle works, also making glass ware for chemical laboratories, pp. 71, 77, Rainfall, p. 56.

Barwick in Elmet (3457), 6½ m. E.N.E. of Leeds. The nave, aisles, and W. tower of the church are Perp. The chancel has a Norm. window. There are fragments of pre-Conquest cross shafts. Here are remarkable earthworks of a Norm. castle.

Bingley

Batley (36,389), a municipal borough, 7 m. S.W. of Leeds. With Morley it returns one M.P. All Saints' Church is chiefly Perp. but partly Dec. The town is a centre of woollen and shoddy manufacture, and makes carpets and rugs and machinery, p. 69. Coal is mined near, p. 82.

Bawtry (1098), a quiet market town on the Idle, close to the Nottinghamshire border, on the Roman Ermine Street, p. 100 ; and the Great North Road, p. 138. The church is Trans. Norm. and E.E. Climate, pp. 55, 56.

Bentham (2476), High Bentham is 1 m. E. of Lower Bentham, and 3 m. S.W. of Ingleton. Lower Bentham parish church tower is Perp. In Lower Bentham are a cotton mill and silk mill ; in High Bentham, canvas hose and belting are made.

Bingley (15,801), is a market town, situated on the terminal moraine of a vanished glacier, on the left bank of the Aire, at 300 ft. above sea-level. It is chiefly known for its manufactories of alpaca, worsted, dress goods, and serge. In the church is a hollowed stone, with three lines of runes.

Birkin (147), 6½ m. S.W. of Selby, has a Norm. church, p. 109.

Birstall (7116), 3 m. N.N.W. of Dewsbury. The lower part of the tower of the church is Norm., the upper part Perp. Woollens, shoddy, carpets and rugs, and woollen machinery are manufactured here. The Elizabethan Oakwell Hall, 1 m. N., is the original of " Field Head " in Charlotte Brontë's *Shirley*. Sculpture, p. 105. Priestley, p. 156.

Bolton Abbey, ruins, 5 m. N.W. by N. of Ilkley, pp. 120, 91.

Bolton by Bowland (635), 5½ m. N.E. by N. of Clitheroe. Hall, p. 131. Church, Perp., p. 112.

Bolton Percy (246), 3 m. E.S.E. of Tadcaster. Church, Perp., p. 112.

Bolton upon Dearne (8670), parish, including Goldthorpe. Colliery. The church is Perp.

Boroughbridge (842), an old-fashioned market town on the Ure. Megalithic monoliths, p. 96. Battle, p. 91.

Boston Spa (1325), inland watering-place, 4 m. W.N.W. of Tadcaster, named from a saline spring discovered in 1744.

Bowling, village in Bradford county borough. Hall, p. 132. Ironworks, p. 72.

Bradfield, eccl. parish (1433), village, 6 m. W.N.W. of Sheffield. Handsome Perp. church. There are two earthworks, Bailey Hill and Castle Hill, which may be Norm. Here are four reservoirs of the Sheffield Corporation. Dale Dyke reservoir burst in 1864, and 244 persons were drowned.

Bradford (288,458), is a city, county borough, and since 1919 the seat of a Bishopric. It returns four members to Parliament. It lies in a hollow, the valley of the Bradford Beck, a tributary of the Aire. The parish church is chiefly Dec. with a Perp. W. tower, and contains a fragment of a pre-Conquest cross shaft. The Exchange, in the Venetian Gothic style, was opened in 1867 ; the Town Hall, also Gothic, with a lofty tower, was completed in 1873. The Cartwright Memorial Hall (1904) contains an art gallery and museum. For history see p. 94. Industries, p. 68. In the suburb of Thornton were born the Brontë sisters. Climate, p. 55.

Bramham cum Oglethorpe (1013), on the Roman Road from Tadcaster to Ilkley, and on the Great North Road. Here is Bramham College. Battle on Bramham Moor, 1408, p. 92.

Brayton (317), 1½ m. S.W. of Selby. Church partly Norm. p. 109.

Brighouse (20,843), a municipal borough on the left bank of the Calder, at 200 ft. above sea-level, and on the Calder and Hebble Canal. It has manufactories of woollens, carpets, and machinery, and soap, cotton-spinning, and silk-spinning mills, p. 71.

Browsholme Hall (*pron.* " Brooslem "), Jacobean residence, 4¼ m. N.W. of Clitheroe, p. 132.

Calverley (2998), 4 m. N.E. of Bradford. The church of St Wilfrid is chiefly Dec., with a Perp. tower, but has Norm. remains above the nave arcade. There are woollen manufactories and stone quarries here.

Campsall (312), 1¼ m. W. of Askern. Church chiefly Trans. Norm., with beautiful rood screen, p. 111.

Castleford (23,090) lies on the right bank of the Aire, at 56 ft. above sea-level, just below the confluence of the Calder. It is the site of the Roman station, Legiolium, p. 100. Roman coins and urns have been found here. It is now known for manufactories of glass bottles and earthenware jars, chemical works, and neighbouring collieries.

Castle Shaw, hamlet, 1½ m. N.N.E. of Delph. Here is a Roman camp, p. 103.

Cawood (955), an old-fashioned village, 1 m. below the confluence of Wharfe and Ouse ; has remains of a castle, pp. 131, 96. The church is partly E.E.

Church Fenton (581), 2 m. N.N.E. of Sherburn. The church is chiefly E.E. with some Dec. and Perp. windows.

Clayton (4863), 3 m. W. by S. of Bradford, has manufactories of alpaca, mohair, and Bradford goods, also quarries and a colliery.

Cleckheaton (12,866), 5 m. S.S.E. of Bradford. Its manufactures consist of woollen cloths, flannels, blankets, worsted yarn, woollen machinery, wire card clothing,[1] and chemicals.

Clifton (2258), 1 m. E. of Brighouse. It has cotton and woollen card and wire manufactures. There is a hamlet called Clifton, 1½ m. S.S.E. of Conisbrough, and another 1½ m. N.N.W. of Otley.

Conisbrough (11,059), a very ancient place on a hillside, sloping to the south bank of the Don, has neighbouring colliery, but interesting Norm. castle, p. 126 ; and a church, partly Trans. Norm., p. 111.   Geology, p. 36.

Cudworth (6824), 3 m. N.E. of Barnsley.   Here are stone quarries and collieries.

Darfield (5427), a colliery village 5 m. E.S.E. of Barnsley. The interesting church is in part Dec., p. 111.

Darrington (514), 2 m. S.E. of Pontefract.   Church shows Trans. Norm., E.E., Dec., and Perp. work.

Darton (5941), 3 m. N.W. of Barnsley, has an interesting Perp. church.   Nail-making and coal mining are carried on here, p. 78.

Denaby (5060), 1 to 2 m. W. of Conisbrough.   Here is a large colliery and a manufactory of explosives.

Dewsbury (53,351) is on the Calder, and is a county borough, returning one M.P.   The church is partly E.E., and contains numerous fragments of Anglo-Sax. sculpture. Industries, pp. 68, 70, 71.

---

[1] The covering of the cylinders of a wool-carding engine, set with wire teeth.

Doncaster (30,516) is a municipal borough on the Don. It is the site of the Roman Danum on Ermine Street, p. 100, and is on the Great North Road, p. 138. The celebrated racecourse belongs to the Corporation. There are here, manufactories of canvas, sacks and ropes, agricultural implements, electric and other motors. In addition to iron foundries and breweries, there are the locomotive and carriage works of the Great Northern Railway. History, p. 93. Friaries, p. 121. Rainfall, p. 56. Wooden houses, p. 130.

Drax (419), 4½ m. W.N.W. of Goole. The church is Trans. Norm., E.E., Dec. and Perp. Of the priory, p. 120, nothing remains but some E.E. fragments.

Drighlington (4126), 5 m. S.E. of Bradford. Worsted spinning and woollen manufactures are the chief occupations.

Ecclesfield (eccl. par., 4671), 4 m. N. of Sheffield. The church is Perp., p. 112. An alien Benedictine priory, belonging to the monastery of St Wandrille, near Rouen, was founded here soon after the Conquest. The thirteenth-century priory chapel has been restored as a curate's residence. Cutlery, files, and nails are manufactured here, and at Thorncliffe, 2 m. N., are three blast furnaces and a colliery.

Elland (10,676), on the S. bank of the Calder, and on the Calder and Hebble Navigation. The church has interesting stained glass. Woollen goods, blankets, flannel, shoddy, and wire card clothing are made here. There are several cotton-spinning and doubling mills.

Elslack (85), 4 m. W.S.W. of Skipton. Here is Burwen Castle, a Roman camp, p. 102.

Farsley (5993), 4 m. E.N.E. of Bradford, has wool-spinning and cloth-weaving mills.

Featherstone (9167), 2 m. W. of Pontefract, is a colliery village. The church contains bells of 1146, and a font of some interest.

Fishlake (475), 2 m. W. of Thorne. The church has a Trans. Norm. S. doorway, a nave of similar or E.E. date, some Dec. windows, and a Perp. W. tower.

Fountains Abbey, ruins, 3 m. S.W. of Ripon, pp. 117, 82, 91.

Fountains Hall is perhaps the most charming old hall in the riding, p. 134.

Giggleswick (946), ½ m. W.N.W. of Settle, has a Perp. church, an ebbing and flowing well, and a Grammar School, founded about 1500, which is one of the two principal public boarding-schools of the West Riding, Sedbergh being the other. Climate, p. 55. Geology, p. 38. Paley, 154.

Golcar (10,110), 3 m. W. by S. of Huddersfield, has woollen manufactories.

Goole (20,332), see chapter on shipping and trade, p. 85.

Guiseley (4925), 2 m. S. by W. of Otley. In the church, the south arcade of the nave is Norm., and there is an E.E. chapel. Chief manufactures, woollen cloths, boots and shoes.

Halifax (101,553), is a county borough, returning one M.P. It is a very hilly town, situated on the Hebble, a tributary of the Calder and on a branch canal. The Perp. parish church is the most interesting building, p. 112. The Town Hall, designed by Sir C. Barry, in the Palladian Renaissance style, was completed in 1863. For local industries, see pp. 66, 68, 71, 77. Natives, pp. 154, 158, 162. Climate, p. 55.

Handsworth (14,198), 3½ m. E.S.E. of Sheffield. The church has some Trans. Norm. features. In the parish are collieries, quarries, and nurseries, p. 152.

Harewood (590), 7 m. N. of Leeds, has a Perp. church, p. 112 ; a ruined castle, p. 128; and Harewood House, p. 135.

Harrogate (33,703), is a municipal borough, fashionable spa and inland watering-place, 13 m. N. of Leeds. The natural mineral waters may be divided into two groups— the Sulphur Group and the Saline Chalybeate and Iron Group.   Climate, p. 55.

Hatfield (1751), 3 m. S.W. of Thorne, has a church with a Trans. Norm. nave and doorways, some Dec. windows, and Perp. chancel, transepts, and central tower.   Battle, p. 89.

Haworth (6505) on a hillside, 3 m. S.S.W. of Keighley, is best known as the former place of residence of the novelist, Charlotte Brontë.   There is a Brontë museum.   Worsted, alpaca, mohair, and Bradford goods are manufactured here.

Healaugh (217), 6 m. E. of Wetherby.   The church is partly Trans. Norm. p. 111.   Remains of Priory 1½ m. S.W.

Heath Old Hall, Elizabethan residence, 1 m. E. of Wakefield, p. 132.

Hebden Bridge (7172), lies in a deep valley at the confluence of the Hebden Water and Calder.   There is attractive wooded scenery at Hardcastle Crags, 3 m. N.W.   In the town are numerous cotton goods manufactories, wholesale clothiers, dye works, etc.

Heckmondwike (9016), 2 m. N.W. of Dewsbury.   Manufactures blankets, shoddy, woollen and worsted goods, carpets and rugs, and wire card clothing.   Native, p. 162.

Hemsworth (10,173), 7½ m. S.E. of Wakefield, is a colliery village, containing Holgate's Hospital, founded by Robert Holgate, archbishop of York, in 1555.

Holmfirth (eccl. par., 4960 ; urban district, 9865), 5½ m. S. of Huddersfield, manufactures woollen, worsted and shoddy. The Bilberry Reservoir, 2½ m. W.S.W., burst in 1852, and 81 lives were lost.

Hoyland Nether (14,638), including Hoyland Common, Upper Hoyland, and Elsecar, 4 m. S.S.E. of Barnsley, has collieries, and brick and tile works.

Hubberholme, 4½ m. N.N.W. of Kettlewell. Church has a rood-loft of 1558.

Huddersfield (107,821) is a county borough, with one M.P. It is situated at 200 to 400 ft. on the left bank of the Colne, a tributary of the Calder, and on the Huddersfield Canal. The chief public buildings are the Town Hall, Municipal Offices, Borough Court, Free Library, and Art Gallery. There is a Technical School. Huddersfield College is a higher grade school. The Literary and Scientific Society has a museum. Industries, pp. 68, 71, 77, Venn, p. 154. Climate p. 55.

Ilkley (7922), a popular inland watering-place, situated at 300 ft. on the right bank of the Wharfe. It is the site of the Roman Olicana, p. 101. Roman inscriptions, tomb stone, altar, amphora, pottery, and coins have been found (see museum). The church is Perp. Sax. crosses, p. 105. Prehistoric remains, p. 96.

Ingleton (1672), 9½ m. N.W. of Settle, is interesting for the neighbouring scenery and geology, pp. 22, 24-26, 36, 39, 81. It lies at the foot of **Ingleborough,** pp. 9, 23, 28, 32, 38, 44, 49, 99.

Keighley, *pron.* " Keethley " (43,487), a municipal borough, situated at 324 ft. in a deep valley near the confluence of Worth and Aire. It has a technical school and good museum. Industries, pp. 70, 77.

Kippax (4075), 5½ m. N.W. by N. of Pontefract, a colliery village, with a Norm. church.

Kirk Hammerton (371), 9 m. W. by N. of York.   The church is partly Sax., p. 108.

Kirklees Priory, ruins, 2 m. E.S.E. of Brighouse, p. 120.

Kirkstall Abbey, ruins, 3 m. N.W. of Leeds, p. 119.

Kirk Hammerton Church

(*Saxon tower, nave and chancel ; modern portion to left of tower*)

Knaresborough (5315) is most picturesquely situated on the summit and steep slopes of a cliff of Magnesian Limestone, on the left bank of the Nidd.   It has a petrifying spring called the Dropping Well, and was the scene of the murder by the schoolmaster, Eugene Aram, notorious owing to Bulwer Lytton's novel.   Church, p. 111.   Castle ruins, p. 127.   Industry, p. 71.   History, pp. 95, 159.

Knaresborough

Knottingley (6680) on the Aire and the Knottingley and Goole Canal, has glass bottle works, chemical works, breweries, shipbuilding yards, and limekilns.

Laughton-en-le-Morthen (1859), 6 m. E.S.E. of Rotherham. The church is partly Sax. and Trans. Norm., p. 109.   Earthworks, p. 124.

Leathley (156), 2 m. N.E. of Otley.   Church has a Norm. tower.

Ledsham (362) 5 m. N. of Pontefract.   Church is partly Saxon, p. 109.

Leeds (445,550 ; in 1801 it was 53,162), a city and county borough on both banks of the Aire, returning six members to Parliament.   In the parish church is a pre-Conquest cross, p. 105.   St John's Church, built 1634, has Carolean Renaissance screen, etc.   The Town Hall (1858) is in Renaissance style.   There is a city Art Gallery.   The Philosophical Hall and Museum contains natural history, geological, and archæological collections.   In 1875 the Yorkshire College was established, and in 1887 became a college of the federal Victoria University.   In 1904 it was incorporated as the University of Leeds.   History of city, pp. 89, 94.   Industries, pp. 66, 68, 71, 77.   Climate, p. 55.   Natives, pp. 159, 160, 161, 162.

Linthwaite (8961), 3 m. W.S.W. of Huddersfield, has woollen and worsted manufactories.

Liversedge (14,658), 3 m. W.N.W. of Dewsbury, manufactures blankets, woollens, worsted, carpets, and woollen machinery, and has cotton-spinning mills.

Markenfield Hall, residence of 1310, 3 m. S.S.W. of Ripon, p. 131.

Marsden (5757), 7 m. S.W. by W. of Huddersfield, manufactures woollens and worsted.

Marston Moor, 7 m. W. of York, battle, 1644, p. 95.

Meltham (5159), 5 m. S.W. by S. of Huddersfield, manufactures woollens, shawls, shoddy, and sewing cotton, and spins cotton and silk.

Methley (4327), 5 m. N.E. of Wakefield. The church is chiefly Dec., but the Waterton Chantry is Perp., pp. 111, 112.

Mexborough (14,401), a colliery town, with glass bottle manufactories. Castle Hill is an earthwork, probably Norm. or pre-Norm., p. 124.

Mirfield (11,712), 4 m. E.N.E. of Huddersfield, manufactures blankets, shoddy and woollens, and woollen machinery, and has cotton-spinning mills. Castle Hall Hill is probably a Norm. motte, p. 124.

Monk Bretton Priory, ruins, 1½ m. E. of Barnsley, p. 116.

Monk Fryston (522), 6 m. N.E. of Pontefract. Church has Saxon tower.

Morley (24,282), a municipal borough, 4 m. S.W. by S. of Leeds. With Batley it returns one M.P. Serge, shoddy, tweed, woollens and worsted, woollen machinery, and miners' safety lamps are manufactured here, pp. 69, 82.

Normanton (15,032) is an important junction of the North-Eastern, Lancashire and Yorkshire, and Midland Railways. The station is said to stand within the moat of a Roman camp. There are collieries near, p. 101.

Nun Monkton (232), 7 m. N.W. by W. of York. Priory church, p. 117.

Ossett (14,078), a municipal borough, 3½ m. W. of Wakefield, manufactures shoddy, woollens, worsted, and woollen machinery.

Otley (9844), on the south bank of the Wharfe, at the foot of Otley Chevin. The church has a Trans. Norm. doorway, and contains fragments of pre-Conquest crosses,

M

p. 105.   In the town are several printers' engineering firms, a worsted spinning mill, rope and twine, and paper manufactories.

Outwood (8459), 2 m. N. of Wakefield.   In the parisn is a colliery.

Pateley Bridge (eccl. par., 2492), the terminus of the Pateley Bridge Branch, North-Eastern Railway, and of the Nidd Valley Light Railway of Bradford Corporation. The neighbouring scenery of Raven's Gill, Guy's Cliffe, and Brimham Rocks, pp. 33, 34, is worth seeing.

Penistone (3408), 12 m. N.W. by N. of Sheffield.   The church is chiefly Perp., p. 112.   Here are steel works.

Pontefract (8647), a municipal borough, 8 m. E. of Wakefield.   The Roman Ermine Street passed 1 m. W. of Pontefract.   There were formerly two priories and five medieval hospitals in Pontefract.   There are ruins of an Elizabethan mansion called Newhall, and of the celebrated castle, p. 127.   Liquorice is cultivated, p. 65.   Natives, p. 159.   History, pp. 91, 96.

Pudsey (14,023), a municipal borough, 5 m. W. of Leeds, has woollen and worsted, carpet and rug manufactories.

Queensbury (6125), 4 m. W.S.W. of Bradford, has manufactures of alpaca, mohair, worsted, and dress goods.

Rawmarsh (17,185), 2 m. N. of Rotherham.   In the parish are iron works (blast furnaces at Parkgate), steel and file works, collieries, and brickyards.

Ripon (8218), a city, municipal borough and seat of a bishopric, is situated between the Ure and the Skell.   There were three medieval hospitals in Ripon, of which two interesting chapels, that of St Mary Magdalene's Hospital and ruins of St Anne's Hospital still exist.   The chief feature of Ripon is its Cathedral, p. 122.   Fountains Abbey lies 3 m. S.W.   Natives, p. 159.

Roche Abbey, ruins, 7½ m. E.S.E. of Rotherham, p. 120.

Rotherham (62,483), a county borough at the confluence of the Rother and Don, has a beautiful Perp. church, p. 112. and a neglected fifteenth-century chantry chapel on the bridge over the Don. In the Clifton Park Museum are Roman remains from Templeborough. In the borough are iron, steel, stove-grate, and glass works, potteries, and collieries. It returns one M.P. pp. 77, 154.

Rothwell (8567), 4 m. S.E. of Leeds, a colliery village. The church has a Perp. tower, a font of Charles II date, and some fragments of late Anglo-Sax. or Norm. sculpture.

Royston (6237), 3¼ m. N.N.E. of Barnsley, a colliery village. The church is Dec. and Perp., p. 111.

Saddleworth (urban district, 12,603 ; eccl. par., 3166), village at 800 to 1000 ft. above sea-level, 5 m. E. by N. of Oldham. The villages of the district are engaged in woollen and shawl manufacture, and calico printing. At Castle Shaw is a Roman camp, p. 103.

Saltaire, near Shipley, p. 69.

Sandal Castle, ruins, 1¼ m. S. of Wakefield, pp. 96, 127.

Sawley Abbey, ruins, 3½ m. N.E. of Clitheroe, p. 120.

Sedbergh (2405), at the foot of Howgill Fells, 8½ m. E. of Kendal. The Church has a Trans. Norm. nave, Perp. windows, and tower. Sedbergh School, founded between 1523 and 1525, is one of the chief public boarding-schools in the riding, and has about 220 boys. Geology, pp. 28, 81.

Selby (9048), on the Ouse, which is crossed by a wooden bridge, is connected by canal with the Aire, and is an important railway junction. Rope and twine making, mustard manufacture, malting, and shipbuilding are the chief industries, but the glory of the town is the abbey church, p. 115. History, pp. 94, 137.

Settle (2583), on the left bank of the Ribble, has a cotton spinning mill and tannery, and is noteworthy for attractive limestone scenery. Victoria Cave, p. 46.

M *

Sharlston (2619), 4 m. E.S.E. of Wakefield.   One mile N. of the village a Roman road runs W. to E., connecting with Ermine Street.   New Sharlston is a colliery village.

Sheffield (459,916 ; in 1801, 45,755), the most populous town in Yorkshire, is a city, county borough, and since 1914, the seat of a bishopric.   It returns seven members to Parliament.   Situated at the confluence of the Don, the Loxley, the Rivelin, the Porter, and the Sheaf, it is a very hilly town, amidst beautiful scenery.   The Firth College was founded in 1879, and in 1905, received a charter as the University of Sheffield.   The Weston Park Museum, with Mappin Art Gallery, and the Ruskin Museum, are of interest.   The Company of Cutlers of Hallamshire, incorporated in 1624, registers trade marks, and performs other useful functions.   Castle, p. 127.   Cathedral, p. 112. Industries, pp. 72-77.   Climate, pp. 55, 57.   Natives, pp. 156, 159, 162.

Sherburn-in-Elmet (1734), 6 m. S. of Tadcaster, on the Great North Road.   The church, p. 111, contains numerous fragments of pre-Conquest crosses.   Æthelstan, and afterwards the archbishops of York, had a palace here, the foundations of which exist in a field called the Hall Garth.

Shipley (including **Saltaire**, 27,706), at the junction of the Bradford Beck with the Aire.   Here are factories of worsted, serge, blankets, and worsted machinery.   Saltaire, p. 69.

Silkstone (1591), 3½ m. W. of Barnsley.   Church Perp., p. 112.   Coal mining, p. 79.

Skipton (12,977), 8 m. N.W. by N. of Keighley, has manufactories of cotton goods and sewing cotton.   Castle, pp. 96, 127, 153.   Church Dec. and Perp., p. 112.   Rainfall, p. 56.

Snaith and Cowick (1619), 7 m. W. of Goole.   Church nave chiefly Perp., p. 111.

Sowerby Bridge (11,350) lies on both sides of the Calder. Here the Calder and Hebble Navigation connects with the

Rochdale Canal. Manufactories of woollen and worsted, carpets, cotton yarn, and textile machinery.

Spenborough, urban district, includes Cleckheaton *q.v.*, Gomersal (3796), and Liversedge, *q.v.*

Spofforth Castle, ruins, 4½ m. S.E. of Harrogate, p. 130.

Springhead (5051), 2½ m. E. of Oldham, has cotton spinning mills.

Stanley (5127), 2½ m. N.N.E. of Wakefield, has jam factory and collieries.

Stocksbridge (7086), 8½ m. N.W. of Sheffield, has steel works.

Swinsty Hall, Elizabethan residence, 5 m. N. of Otley, p. 132.

Swinton (13,654), 4½ m. N.N.E. of Rotherham, has glass bottle works, railway waggon works, stove and grate manufactories. Rockingham faience, p. 77.

Syningthwaite Priory, 3½ m. E. of Wetherby, p. 120.

Tadcaster (East and West together, 3399), is situated at 40 ft. above sea level on both banks of the Wharfe, and was the Roman station of Calcaria, p. 100. The Castle Hill appears to have been the site of a Norm. castle. In 1341 there were two breweries in the town, and at the present day it is best known for its breweries. Battle, 1642, p. 94.

Temple Newsam (including Whitkirk, 3346), 2 m. E.S.E. of Leeds. The Manor House is Jacobean, p. 135.

Thorne (5290), 7 m. S.W. by S. of Goole. The church is Trans. Norm., Dec., and Perp. Peat moss litter gathering and warping find employment.

Thornhill (11,303), civil parish in Dewsbury. The interesting church of St Michael has a Perp. tower, modern nave, fine ancient stained glass, and monuments of the Savilles, p. 105. Lees Hall, p. 130.

Thorpe Salvin (373), 11 m. E.S.E. of Sheffield. Church partly Trans. Norm., p. 109.

Throapham (St John's with, 95), 1 m. from Laughton-en-le-Morthen, and 7 m. E.S.E. of Rotherham.   St John's Church is Trans. Norm. and Perp.

Tickhill (1806), 5 m. E. of Rotherham.   Castle ruins, pp. 95, 126.   Perp. church, p. 112.   Friary, p. 121.

Todmorden (25,404), a municipal borough in the deep and narrow Calder Gap, pp. 8, 20.   Its manufactures are cotton goods, fustians, cotton yarns, cotton machinery, and fancy dress goods.

Towton (95), 2½ m. S. of Tadcaster, battle, 1461, p. 92.

Treeton (1859), colliery village, 3 m. S. of Rotherham. The lower part of the tower, chancel arch, and inner doorway of porch of the church are Trans. Norm.

Wakefield (51,511), a city, parliamentary and county borough, seat since 1888 of a bishopric, shire town of the administrative county of the West Riding, is situated on the Calder, pp. 130, 156, 158, 159.   Cathedral, p. 112. Chapel on the bridge, p. 111.   Battle, pp. 92, 94.   Industries, pp. 66, 68, 70, 77.   Climate, p. 55.

Wath-upon-Dearne (7331), 5 m. N. of Rotherham, a colliery village.   The church is Trans. Norm., Dec., and Perp.

Wentworth Woodhouse, residence, 3½ m. N.W. of Rotherham. pp. 135, 132.

Weston Hall, Elizabethan, 2 m. N.W. of Otley, p. 132.

Wincobank, camp, near Sheffield, p. 99.

Wombwell (17,536), 4 m. S.E. by E. of Barnsley, colliery town.

Woodsome Hall, Tudor residence, 2½ m. S.E. of Huddersfield, p. 132.

Worsborough (12,750), 2¼ m. S. of Barnsley.   The church has Trans. Norm. work and contains curious recumbent wooden effigy of Roger Rockley (*d.* 1534).

Yeadon (7440), 3 m. S. of Otley, manufactures woollens, shawls, tweeds, and dress goods.   Turnpike riots, p. 137.

The Shambles, York

York (82,282), the county town of Yorkshire, a parliamentary and county borough and archiepiscopal city, is situated on both banks of the Ouse. During the Roman occupation it was, as Eboracum, the first or second city of Britain ; in Anglian and Danish times the capital of the kingdom of Deira; afterwards for centuries the second city of England. It lies on Ermine Street and on the Great North Road, and is an important railway centre. Its chief features have been already described, viz., the Roman fort, p. 100 ; the castle, p. 124 ; the medieval walls and gates, p. 128 ; the Cathedral, p. 121 ; St Mary's Abbey, p. 116 ; Norm. doorways, p. 109 ; half-timber houses, p. 130 ; history, pp. 88-95 ; industries, pp. 66, 77. In the Guild Hall is a handsome old hall in Perp. style, erected in 1446. St William's College, opposite the Minster, founded in 1461 as a residence for Chantry priests, has a Perp. entrance doorway, but interiorly is chiefly Jacobean. The King's Manor, E. of St Mary's Abbey Church, was originally built between 1485 and 1495 as the Abbot's House and converted into a palace for the Lord President of the North, and is occupied as a school; it has some interesting seventeenth century doorways. The Museum of the Yorkshire Philosophical Society contains antiquities and natural history collections. There are many interesting old churches. York is a county of itself and a quarter sessions borough. The assizes of the N. and E. Ridings and the City of York are held at York. In the Ministry of Health official list and for Board of Control and Mental Deficiency purposes, also in Agricultural Statistics, York is treated as being in the E. Riding. The Lord Lieutenant of the W. Riding is also Lord Lieutenant of York, and York is associated with that Riding for military purposes such as the Territorial Association. York returns one member to Parliament. Etymology of name, p. 1. Geology, p. 41. Climate, p. 55. Birthplace of Etty, Flaxman, and Barnby, p. 162.

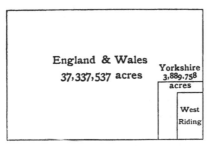

Fig. 1. Area of the West Riding (1,773,529 acres) compared with that of Yorkshire, and of England and Wales

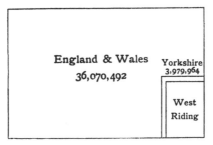

Fig. 2. Population of the West Riding (3,045,377) compared with that of Yorkshire, and of England and Wales at the last Census

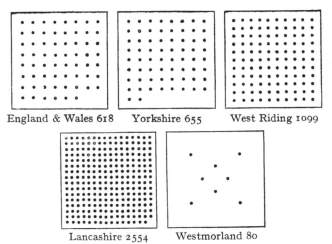

England & Wales 618   Yorkshire 655   West Riding 1099

Lancashire 2554   Westmorland 80

Fig. 3. Comparative Density of the Population
to the square mile at the last Census
(*Each dot represents ten persons*)

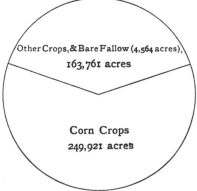

Other Crops, & Bare Fallow (4,564 acres),
163,761 acres

Corn Crops
249,921 acres

Fig. 4. Area under Cereals compared with that of other
Farmed Land in the West Riding

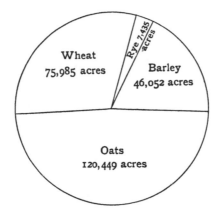

Fig. 5. Proportionate Areas of Chief
Cereals in the West Riding

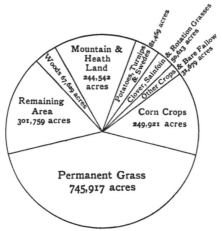

Fig. 6. Proportionate Areas of Cultivated
and Uncultivated Land in the West
Riding

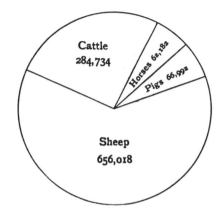

Fig. 7.   Proportionate numbers of Live
Stock in the West Riding

Milton Keynes UK
Ingram Content Group UK Ltd.
UKHW041520181024
449640UK00009B/94